Mastering ABP Framework

Build maintainable .NET solutions by implementing software development best practices

Halil İbrahim Kalkan

ABP IS AN IMPRINT OF PACKT PUBLISHING

Mastering ABP Framework

Group Product Manager: Pavan Ramchandani

Publishing Product Manager: Aaron Tanna

Senior Editor: Hayden Edwards

Content Development Editor: Rashi Dubey

Technical Editor: Simran Udasi

Copy Editor: Safis Editing

Project Coordinator: Rashika Ba

Proofreader: Safis Editing

Indexer: Tejal Daruwale Soni

Production Designer: Alishon Mendonca

Marketing Coordinator: Anamika Singh

First published: January 2022

Production reference: 1280122

Published by Packt Publishing Ltd.

Livery Place

35 Livery Street

Birmingham

B3 2PB, UK.

ISBN 978-1-80107-924-2

www.packt.com

To my wife, Gözde, for always trusting me and being with me, and to my kids, Tuana and Mete.

– Halil İbrahim Kalkan

Foreword

I have known Halil İbrahim Kalkan for almost 20 years and we have been working together for more than 10 years. Halil is one of the greatest developers/software architects I have ever worked with. He likes solving common software problems and tries to create solutions that other developers can use. With this passion, he created ASP.NET Boilerplate Framework back in 2013 and then included me in his journey of framework development. After ASP.NET Boilerplate became very popular in the .NET world, Halil decided to re-write it with the name ABP Framework. ABP Framework is a complete infrastructure for creating modern web applications by following software development best practices and conventions.

In this book, Halil will show you how easy it is to build robust, maintainable, and scalable software solutions using ABP Framework. Developing modular software is every developer's dream, but it is really hard to do. Halil will also show you how to create modular software easily using ABP Framework.

You will also learn how to work with Entity Framework Core and MongoDB to develop the data access layer for your application, how to build UIs with ASP.NET Core MVC (Razor Pages) and Blazor, and how ABP Framework seamlessly supports multi-tenancy.

To develop robust software, testing the application is very important. Halil will guide you on how to write unit and integration tests using ABP Framework easily.

By the end of this book, you will be able to create a complete web solution that is easy to develop, maintain, and test.

İsmail ÇAĞDAŞ

Co-Founder – Volosoft

Contributors

About the author

Halil İbrahim Kalkan is a computer engineer who loves building reusable libraries, creating distributed solutions, and working on software architectures. He is an expert in domain-driven design, multi-tenancy, modularity, and microservice architecture. Halil has been building software since 1997 (when he was 14) and working as a professional since 2007. He has a lot of articles and talks on software development. He is a very active open source contributor and has built many projects based on web and Microsoft technologies. Halil is currently leading the open source ABP Framework, which provides a complete architectural solution for building .NET applications.

I want to thank the people who were close to me and supported me while writing this book, especially my wife, Gözde, for her understanding and patience with me spending my all weekends completing this book, and my mother, Fatma, for always supporting me. I also thank Aaron Tanna, who came to me with this book offer; Hayden Edwards, Rashi Dubey, and Lee Richardson, for their excellent reviews and editorial support; and Berkan Şaşmaz, Engincan Veske, Armağan Ünlü, Melis Platin, and Enis Necipoğlu, for their help in preparing the EventHub reference application. This book could not have existed without your help.

About the reviewer

Lee Richardson is a Microsoft MVP and a prolific writer, speaker, and YouTuber on .NET and open source topics. He frequently covers ABP content on his popular "Code Hour" YouTube channel, his blog, and on CodeProject. Throughout his two decades of software development consulting in the Washington DC area, he has spoken scores of times at code camps, conferences, and user groups. He currently works as a Solution Samurai at InfernoRed Technologies. He is active on Twitter, where you can reach him @lprichar.

Table of Contents

3

Step-By-Step Application Development

4

Understanding the Reference Solution

Part 2: Fundamentals of ABP Framework

5

Exploring the ASP.NET Core and ABP Infrastructure

6

Working with the Data Access Infrastructure

7

Exploring Cross-Cutting Concerns

8

Using the Features and Services of ABP

Part 3: Implementing Domain–Driven Design

9
Understanding Domain-Driven Design

10
DDD – The Domain Layer

11
DDD – The Application Layer

Part 4: User Interface and API Development

12
Working with MVC/Razor Pages

13
Working with the Blazor WebAssembly UI

14
Building HTTP APIs and Real-Time Services

Part 5: Miscellaneous

15
Working with Modularity

16
Implementing Multi-Tenancy

17
Building Automated Tests

Index

Other Books You May Enjoy

Preface

ABP Framework is a complete infrastructure for creating modern web applications by following software development best practices and conventions. ABP provides a high-level framework and ecosystem to help you implement the Don't Repeat Yourself (DRY) principle and focus on your business code.

Written by the creator of ABP Framework, this book will help you to gain a complete understanding of ABP Framework and modern web application development techniques from scratch. With step-by-step explanations of essential concepts and practical examples, you'll understand the requirements of a modern web solution and how ABP Framework makes it enjoyable to develop your own solution. You'll discover the common requirements of enterprise web application development and explore the infrastructure provided by ABP Framework. Throughout the book, you'll get to grips with software development best practices for building maintainable and modular web solutions.

By the end of this book, you'll be able to create a complete web solution that is easy to develop, maintain, and test.

Who this book is for

This book is for web developers who want to learn software architecture and best practices to build maintainable web-based solutions using Microsoft technologies and ABP Framework. Basic knowledge of C# and ASP.NET Core is necessary to get started with this book.

What this book covers

Chapter 1, Modern Software Development and ABP Framework, discusses the common challenges of developing a business application and explains how ABP addresses these challenges.

Chapter 2, Getting Started with ABP Framework, explains how you can create and run a new solution with ABP Framework.

Chapter 3, Step-By-Step Application Development, is the longest chapter of the book and shows application development with ABP Framework with an extensive example application. It is the main chapter for putting everything together. After this chapter, you may not understand all the ABP features but will be able to create your own application with the fundamental ABP concepts. You will understand the big picture here. Then, you will fill in the gaps and learn about all the details in the next chapters.

Chapter 4, Understanding the Reference Solution, explains the architecture and structure of the reference solution, EventHub, that was created as a large example application for the readers of this book. It is suggested that you read this chapter and make the solution work in your environment.

Chapter 5, Exploring the ASP.NET Core and ABP Infrastructure, explains some fundamental concepts such as dependency injection, basic modularity, configuration, and logging. These topics are essential to understanding development with ABP and ASP.NET Core.

Chapter 6, Working with the Data Access Infrastructure, introduces the entity, repository, and unit of work concepts, and shows how to work with Entity Framework Core and MongoDB. You will learn different ways of querying and manipulating data and controlling database transactions.

Chapter 7, Exploring Cross-Cutting Concerns, focuses on three important concerns you will need in your application: authorization, validation, and exception handling. These concerns are implemented in every part of your application. You will learn how to define and use a permission-based authorization system, validate user inputs, and deal with exceptions and exception messages.

Chapter 8, Using the Features and Services of ABP, covers some of the commonly used ABP features such as working with the current user, using the data filtering and audit log systems, caching data, and localizing the user interface.

Chapter 9, Understanding Domain-Driven Design, is the first part of the DDD-related chapters. It starts by defining DDD and structuring a .NET solution based on DDD. You will learn how ABP's startup template has been evolved from DDD's standard four-layered solution model. You will also learn about the building blocks and principles of DDD.

Chapter 10, DDD – The Domain Layer, continues looking at DDD with the domain layer. It begins by explaining the EventHub domain objects, since the examples in this chapter and the next chapter will be based on these objects. You will learn how to design aggregates; implement domain services, repositories, and specifications; and use an event bus to publish domain events.

Chapter 11, DDD – The Application Layer, focuses on the application layer. You will learn the best practices for designing and validating data transfer objects and implementing your application services. You will also find discussions and examples in this chapter to help you understand the responsibilities of the domain and application layers.

Chapter 12, Working with MVC/Razor Pages, covers developing MVC (Razor Pages) applications that generate the HTML on the server side. You will learn about ABP's theming approach and learn about fundamental aspects such as bundling and minification, tag helpers, forms, menus, and modals. You will also learn how to make client-to-server API calls and use the JavaScript utility APIs provided by ABP Framework to show notifications, message boxes, and more.

Chapter 13, Working with the Blazor WebAssembly UI, is similar to the previous chapter and explains UI development with Microsoft's new Blazor SPA framework and ABP Framework. Blazor is a great framework for developers to use their existing .NET skills in the browser. ABP takes things a few steps further by providing built-in solutions for consuming HTTP APIs, implementing theming, and providing utility services to make common UI tasks easier.

Chapter 14, Building HTTP APIs and Real-Time Services, explains how to create API controllers with the classic ASP.NET approach and ABP's auto API controller system, and it discusses when you need to define controllers manually. In this chapter, you will also learn about dynamic and static C# proxies to automate client-to-server API calls from your .NET clients to your ABP-based HTTP services. This chapter also covers using SignalR with ABP Framework.

Chapter 15, Working with Modularity, explains reusable application module development with an example case. We'll create a payment module for the EventHub solution and explain the structure of that module in this chapter. In this way, you will understand how to develop reusable modules and install them in an application.

Chapter 16, Implementing Multi-Tenancy, focuses on another fundamental ABP architecture, multi-tenancy, which is an architectural pattern to build **Software-as-a-Service (SaaS)** solutions. You will see whether multi-tenancy is the right architecture for your solution and learn how to develop your code to be compatible with ABP's multi-tenancy system. This chapter also covers ABP's features system, which is used to define application functionalities as features and assign them to tenants in a multi-tenant solution.

Chapter 17, Building Automated Tests, explains ABP's test infrastructure and how to build unit and integration tests for your applications using **xUnit** as the test framework. You will also learn the basics of automated tests, such as assertions, mocking and replacing services, and dealing with exceptions.

To get the most out of this book

Basic knowledge of C# and ASP.NET Core is necessary to get started with this book.

Software/hardware covered in the book	Operating system requirements
.NET 6.0, ASP.NET Core 6.0	Windows, macOS, or Linux
ABP Framework 5.0	Windows, macOS, or Linux

If you are using the digital version of this book, we advise you to type the code yourself or access the code from the book's GitHub repository (a link is available in the next section). Doing so will help you avoid any potential errors related to the copying and pasting of code.

Download the example code files

You can download the example code files for this book from GitHub at `https://github.com/PacktPublishing/Mastering-ABP-Framework`. If there's an update to the code, it will be updated in the GitHub repository.

We also have other code bundles from our rich catalog of books and videos available at `https://github.com/PacktPublishing/`. Check them out!

Download the color images

We also provide a PDF file that has color images of the screenshots and diagrams used in this book. You can download it here: `https://static.packt-cdn.com/downloads/9781801079242_ColorImages.pdf`

Conventions used

There are a number of text conventions used throughout this book.

`Code in text`: Indicates code words in text, database table names, folder names, filenames, file extensions, pathnames, dummy URLs, user input, and Twitter handles. Here is an example: "If you want to specify the database connection string, you can also pass the `--connection-string` parameter as shown in the following example:"

A block of code is set as follows:

```
"ConnectionStrings": {
  "Default": "Server=(LocalDb)\\
MSSQLLocalDB;Database=ProductManagement;Trusted_
Connection=True"
}
```

Any command-line input or output is written as follows:

```
dotnet tool install -g Volo.Abp.Cli
```

Bold: Indicates a new term, an important word, or words that you see onscreen. For instance, words in menus or dialog boxes appear in **bold**. Here is an example: "ABP Framework provides a pre-built **Application Startup Template**."

> **Tips or Important Notes**
> Appear like this.

Get in touch

Feedback from our readers is always welcome.

General feedback: If you have questions about any aspect of this book, email us at customercare@packtpub.com and mention the book title in the subject of your message.

Errata: Although we have taken every care to ensure the accuracy of our content, mistakes do happen. If you have found a mistake in this book, we would be grateful if you would report this to us. Please visit www.packtpub.com/support/errata and fill in the form.

Piracy: If you come across any illegal copies of our works in any form on the internet, we would be grateful if you would provide us with the location address or website name. Please contact us at copyright@packt.com with a link to the material.

If you are interested in becoming an author: If there is a topic that you have expertise in and you are interested in either writing or contributing to a book, please visit authors.packtpub.com.

Part 1: Introduction

The first part of this book introduces the challenges of modern web application development and explains how ABP solves the problems. It also shows how to create a new solution with ABP Framework and build a fully working, production-ready page to manage an entity. Finally, it explores the EventHub project, a real-world reference solution built with ABP Framework.

In this part, we include the following chapters:

- *Chapter 1, Modern Software Development and ABP Framework*
- *Chapter 2, Getting Started with ABP Framework*
- *Chapter 3, Step-by-Step Application Development*
- *Chapter 4, Understanding the Reference Solution*

1
Modern Software Development and ABP Framework

Building software systems has always been complicated. Especially in these modern times, there are many challenges while creating even a basic business solution. You often find yourself implementing standard non-business requirements and digging into infrastructure problems rather than implementing your business code, which is the actual valuable part of the system you are trying to build.

ABP Framework helps you focus on the code that adds value to the stakeholders by offering a robust software architecture, automating the repetitive details, and providing the necessary infrastructure to help build modern web solutions. It provides an end-to-end, consistent development experience and improves your productivity. ABP gets you and your team up to speed with all the modern software development best practices pre-applied.

This book is the ultimate guide to developing web applications and systems using ABP Framework by following modern software development approaches and best practices.

This first chapter introduces the challenges of building a well-architected enterprise solution and explains how ABP Framework addresses these challenges. I will also explain the purpose and the structure of this book.

In this chapter, we will cover the following topics:

- Challenges of developing an enterprise web solution
- Understanding what ABP Framework offers

Challenges of developing an enterprise web solution

Before digging into ABP Framework, I want to present the challenges of developing a modern enterprise web solution to understand why we need an application framework like ABP Framework. Let's begin with the big picture: architecture.

Setting up the architecture

Before you start to write your code, you need to create a foundation for your solution. This is one of the most challenging phases of building a software system. You have a lot of options and need to make some fundamental decisions. Any decision you make at this stage will likely affect your application for the rest of its lifetime.

There are some common, well-known, system-level architectural patterns, such as **monolithic architecture**, **modular architecture**, and **microservice architecture**. Applying one of these architectures determines how you develop, deploy, and scale your solution and should be decided based on your requirements.

In addition to these system-level patterns, software development models such as **Command and Query Responsibility Segregation (CQRS)**, **Domain-Driven Design (DDD)**, **Layered Architecture**, and **Clean Architecture** determine how your code base is shaped.

Once you decide on your architecture, you should create the fundamental solution structure to start development with that architecture. In this phase, you also need to decide which language, framework, tools, and libraries you will use.

All these decisions need significant experience, so they are ideally done by experienced software architects and developers. However, not all the team members will have the same experience and knowledge level. You need to train them and determine the correct coding standards.

After setting up your architecture and preparing the fundamental solution, your team can start the development process. The next section discusses the common aspects that are repeated by every software solution and how you can avoid repeating them in your development.

Don't repeat yourself!

Don't Repeat Yourself (**DRY**) is a key principle for software development. Computers automate the repetitive tasks of the real world to make people's lives easier. So, why do we repeat ourselves while building software solutions?

Authentication is a very common concern of every software solution – single sign-on, Active Directory integration, token-based authentication, social logins, two-factor authentication, forgot/reset password, email activation, and more. Are most of these requirements are familiar to you? You are not alone! Almost all software projects have more or less similar requirements for authentication. Instead of building all these from scratch, using an existing solution, such as a library or a cloud service, is better. Such pre-built solutions are mature and battle-tested, which is important for security.

Some non-functional requirements, such as exception handling, validation, authorization, caching, audit logging, and database transaction management, are other sources of code repetition. These concerns are called cross-cutting concerns and should be handled in every web request. In a well-architected software solution, these concerns should be handled automatically by conventions in a central place in your code base, or you should have services to make them easier to implement.

When you integrate to third-party systems, such as RabbitMQ and Redis, you typically want to create abstractions and wrappers around the code that interact with these systems. In this way, your business logic is isolated from these infrastructure components. Also, you don't repeat the same connection, retry, exception handling, and logging logic everywhere in your solution.

Having a pre-built infrastructure to automate this repetitive work saves your development time so that you can focus on your business logic. The next section discusses another topic that takes up our time in every business application – the user interface.

Building a UI base

One of the fundamental aspects of an application is its **user interface** (**UI**). An application with an unfashionable and unusable UI would not be as attractive at first glance, even if it has outstanding business value under the hood.

While UI features and requirements vary for every application, some fundamental structures are common. Most applications need basic elements, such as alerts, buttons, cards, form elements, tabs, and data tables. You can use HTML/CSS frameworks such as Bootstrap, Bulma, and Ant Design instead of creating a design system for every application.

Almost every web application has a responsive layout with the main menu, toolbar, header, and footer with custom colors and branding. You will need to determine all these and implement a base UI kit for your application's pages and components. In this way, UI developers can create a consistent UI without dealing with the common structures.

Up to here, I've introduced some common infrastructure requirements, mostly independent from any business application. The next section discusses common business requirements for most enterprise systems.

Implementing common business requirements

While every application and system is unique and their value comes from that uniqueness, every enterprise system has some fundamental supporting requirements.

A permission-based authorization system is one of these fundamental requirements. It is used to control the privileges of users and clients of the application. If you want to implement this yourself, you should create an end-to-end solution with database tables, authorization logic, permission caches, APIs, and UI pages to assign these permissions to your users and check them when needed. However, such a system is pretty generic and can be developed as a shared identity management functionality (a reusable module) and used by multiple applications.

Like identity management, many systems need functionalities such as audit log reporting, tenant and subscription management (for SaaS applications), language management, file uploading and sharing, multi-language management, and time zone management. In addition to the pre-built application functionalities (modules), there may be low-level requirements, such as implementing the soft-delete pattern and storing **Binary Large Object (BLOB)** data in your applications.

All these common requirements can be built from scratch, which can be the only solution for some enterprise systems. However, if these functionalities are not the main value that's provided by your application, you can consider using pre-built modules and libraries where they are available and customize them based on your requirements.

In the next section, you will learn how ABP Framework helps us with the common infrastructure and base requirements that were discussed in this section.

Understanding what ABP Framework offers

ABP Framework offers an opinionated architecture to help you build enterprise software solutions with best practices on top of the .NET and ASP.NET Core platforms. It provides the fundamental infrastructure, production-ready modules, themes, tooling, guides, and documentation to implement that architecture properly and automate the details and repetitive work as much as possible.

In the next few sub-sections, I will explain how ABP does all these, starting with the architecture.

The ABP architecture

I mentioned that ABP offers an opinionated architecture. In other words, it is an opinionated framework. So, I should first explain what an unopinionated framework is and what an opinionated framework is.

As I stated in the *Setting up the architecture* section, preparing a foundation for a software solution requires a lot of decisions; you should decide on the system architecture, development model, techniques, patterns, tools, and libraries to use in your solution.

Unopinionated frameworks, such as ASP.NET Core, don't say much about these decisions and mostly leave it up to you. For example, you can create a layered solution by separating your UI layer from the data access layer, or you can create a single-layered solution by directly accessing the database from your UI pages/views. You can use any library, so long as it is compatible with ASP.NET Core, and you can apply any architectural pattern. Being unopinionated makes ASP.NET Core flexible and usable in different scenarios. However, it assigns the responsibility to you to make all these decisions, set up the right architecture, and prepare your infrastructure to implement that architecture.

I don't mean ASP.NET Core has no opinion at all. It assumes you are building a web application or API based on the HTTP specification. It clearly defines how your UI and API layers should be developed. It also offers some low-level infrastructure components such as dependency injection, caching, and logging (in fact, these components are usable in any .NET application and not specific to ASP.NET Core, but they are mainly developed alongside ASP.NET Core). However, it doesn't say much about how your business code is shaped and which architectural patterns you will use.

ABP Framework, on the other hand, is an opinionated framework. It believes that certain ways of approaching software development are inherently better and thus guide developers down those paths. It has opinions about the architecture, patterns, tools, and libraries you will use in your solution. Though ABP Framework is flexible enough to use different tools and libraries, and change your architectural decisions, you get the best value when you follow its opinions. But don't worry; it provides good, industry-accepted solutions to common architectures to help you build maintainable software solutions with best practices. The decisions it takes will save your time, increase your productivity, and make you focus on your business code rather than infrastructural problems.

In the next few sections, I will introduce the four fundamental architectures ABP stands on.

Domain-driven design

ABP's main goal is to provide a model to build maintainable solutions with clean code principles. ABP offers a layered architecture based on DDD patterns and practices. It provides a layered startup template (see *The startup templates* section), the necessary infrastructure, and guidance for applying that architecture properly.

Since ABP is a software framework, it focuses on the technical implementation of DDD. *Part 3*, *Implementing Domain-Driven Design*, of this book explains the best practices of building a DDD-based solution using ABP Framework.

Modularity

In software development, modularity is a technique that's used to split a system into isolated parts, called modules. The ultimate goal is to reduce complexity, increase reusability, and enable different teams to work on different sets of features in parallel without affecting each other.

Modularity has two main challenges that are simplified with ABP Framework:

- The first challenge is to isolate modules. ASP.NET Core has some features (such as Razor component libraries) to support modular applications. Still, it is very limited because it is an unopinionated framework and has opinions only for the UI and API parts. On the other hand, ABP Framework provides a consistent model and infrastructure to build fully isolated, reusable application modules with its database, domain, application, and UI layers.

- The second challenge of modularity is dealing with how these isolated modules communicate and become a single, unified application at runtime. ABP offers concrete models for common requirements of a modular system, such as sharing a database among modules, communicating between the modules via events or API calls, and installing a module in an application.

ABP provides many pre-built open source application modules that can be used in any application. Some examples include the Identity module, which provides user, role, and Permission Management, and the Account module, which provides login and register pages for your application. Reusing and customizing these modules saves your time. In addition, ABP provides a module startup template to help you build reusable application modules. An example of this can be found in *Chapter 15, Working with Modularity*.

Modularity is great for managing the complexity of a large monolithic system. However, ABP helps you create microservice solutions too.

Microservices

Microservices and distributed architecture is the accepted approach to building scalable software systems. It allows different teams to work on different services and independently version, deploy, and scale their services.

However, building a microservice system has some important challenges in terms of development, deployment, inter-microservice communication, data consistency, monitoring, and more.

Microservice architecture is not a problem that a single software framework can solve. A microservice system is a solution that brings many different disciplines, approaches, technologies, and tools together to solve unique problems. Every microservice system has its requirements and restrictions. Each team has a level of expertise, knowledge, and skills.

ABP Framework was designed to be microservice compatible from the beginning. It provides a distributed event bus for asynchronous communication between microservices with transaction support (as explained in the *Publishing domain events* section of *Chapter 10, DDD – The Domain Layer*). It also provides C# client-side proxies to easily consume the REST APIs of remote services (as explained in the *Consuming HTTP APIs* section of *Chapter 14, Building HTTP APIs and Real-Time Services*).

All of the pre-built ABP application modules are designed so that you can convert them into microservices. ABP also provides a detailed guide (`https://docs.abp.io/en/abp/latest/Best-Practices/Index`) to explain how you can create such microservice-compatible modules. In this way, you can start with a modular monolith, and then convert it into a microservice solution later.

The core ABP team has prepared an open source microservice reference solution built with ABP Framework. It demonstrates how you can create a solution with API Gateways, inter-microservice communication, distributed events, distributed caches, multiple database providers, and multiple UI applications with single sign-on. It also includes the Kubernetes and Helm configurations to run the solution on containers. See `https://github.com/abpframework/eShopOnAbp` to learn all the details about that solution.

The next section introduces the last fundamental architecture that ABP Framework provides out of the box – multi-tenancy.

SaaS/multi-tenancy

Software-as-a-Service (**SaaS**) is a trending approach to building and selling software products. Multi-tenancy is a widely used architectural pattern for building SaaS systems. The following are the typical features of a multi-tenant system:

- Shares the hardware and software resources between tenants.
- Every tenant has users, roles, and permissions.
- Isolates database, cache, and other resources between tenants.
- Can enable/disable application features per tenant.
- Can customize application configurations per tenant.

ABP Framework covers all these requirements and more. It helps you build a multi-tenant system while most of your code base is unaware of multi-tenancy.

Chapter 16, Implementing Multi-Tenancy, explains multi-tenancy and multi-tenant application development with ABP Framework.

So far, I've introduced the fundamental architectural patterns that ABP provides as pre-built solutions. However, ABP also provides startup templates to help you get started with a new solution easily.

The startup templates

When you create a new solution using ASP.NET Core's standard startup templates, you get a single-project solution with minimal dependencies and no layers, which is not so production-ready. You usually spend a considerable amount of time setting up the solution structure to implement your software architecture properly, as well as to install and configure the fundamental tools and libraries.

ABP Framework provides a well-architected, layered, pre-configured, and production-ready startup solution template. The following screenshot shows the initial UI when you directly run the startup template that's created with ABP Framework:

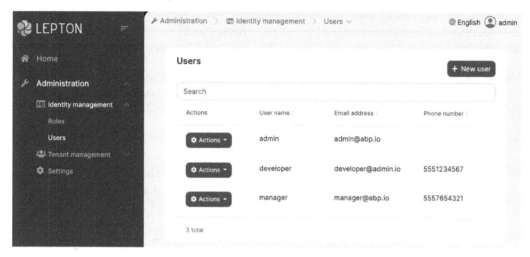

Figure 1.1 – ABP application startup template

Let's talk about this startup template in more detail:

- The solution is layered. It is clear and tells you how to organize your code base.

- Some pre-built modules are already installed, such as the **Account** and **Identity** modules. You have *log in, register, user and role management,* and some other standard functionalities already implemented.

- **Unit test** and **integration test** projects are pre-configured and ready to write your first test code.

- It contains some utility applications to manage your database migrations and consume and test your HTTP APIs.

ABP's application startup template comes with multiple options for the **UI Framework** and the **Database Provider**. You can start with **Angular**, **Blazor**, or **MVC** (**Razor Pages**) options as the UI framework, and use **Entity Framework Core** (with any database management system) or **MongoDB** as the database provider. You will learn how to create a new solution and run it in *Chapter 2, Getting Started with ABP Framework*.

In the next section, I will introduce some of ABP's infrastructure components.

The ABP infrastructure

ABP is based on familiar tools and libraries you already know about. While it is a full-stack application framework, it doesn't introduce a new **Object-Relational Mapper (ORM)** and instead uses Entity Framework Core. Similarly, it uses Serilog, AutoMapper, IdentityServer, and Bootstrap instead of creating similar functionalities itself. It provides a solution that integrates these tools, fills the gaps, and implements common business application requirements.

ABP Framework simplifies exception handling, validation, authorization, caching, audit logging, and database transaction management by automating them by conventions and allowing you to fine-control when you need to. So, you don't repeat yourself for these cross-cutting and common concerns.

ABP is well integrated with IdentityServer for cookie and token-based authentication, as well as single-sign-on. It also provides a detailed, permission-based authorization system to help you control the privileges of the users and clients of the application.

Besides the basics, background jobs, BLOB storage, text templating, audit logging, and localization components provide built-in solutions for common business requirements.

On the UI part, ABP provides a complete UI theming system to help you develop theme-unaware and modular applications and easily install a theme for an application. It also provides tons of features and helpers on the UI side to eliminate repetitive code and increase productivity.

The next section will talk about the community, which is important for an open source project.

The community

When you set up your solution architecture in your company, no one knows your structure except the developers working on it. However, ABP has a large and active community. They are using the same architecture and infrastructure, applying similar best practices, and developing their application similarly. This has a great advantage when you are stuck with an infrastructure problem or want to get an idea or a suggestion for implementing a business problem. It is also easier to understand someone's code in another solution since ABP developers are applying the same or similar patterns.

ABP Framework has been around and growing since 2016. At the end of 2021, it has 7,000+ stars, 220+ contributors, 22,000+ commits, 5,700 closed issues on GitHub, and more than 4,000,000 downloads on NuGet with more than 110+ major and minor releases. I mean, it is a mature, accepted, and trusted open source project.

The core ABP team and the contributors from the community are constantly writing articles, preparing video tutorials, and sharing on the ABP Community website: `https://community.abp.io`. The following screenshot has been taken from the ABP Community website:

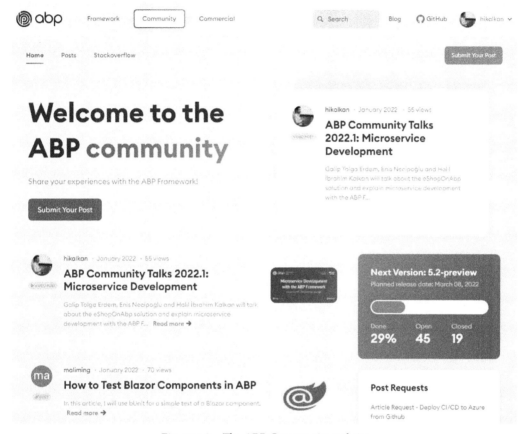

Figure 1.2 – The ABP Community website

Check out the ABP Community website to see what others are doing with ABP Framework and closely follow ABP Framework's development.

Summary

In this chapter, we introduced the problems of building a business solution and explained how ABP provides solutions to these common problems. ABP also increases developer productivity by providing a pre-built architectural solution and the necessary infrastructure to implement that architecture.

By the end of this book, you will be comfortable with ABP Framework and will have learned a lot of best practices and techniques regarding enterprise software development.

In the next chapter, you will learn how to create a new solution using ABP's **command-line interface (CLI)** tool and run it in your development environment.

2
Getting Started with ABP Framework

ABP Framework is distributed as a large set of NuGet and **Node Package Manager (NPM)** packages. It has a modular design so that you can add and use the packages you need to. However, there are also some pre-built solution templates, and you typically want to start with them.

We will see how to prepare our development environment and create solutions using ABP's startup templates. By the end of this chapter, you will have a running solution built with ABP Framework.

This chapter consists of the following topics:

- Installing the ABP CLI
- Creating a new solution
- Running the solution
- Exploring the pre-built modules

Technical requirements

There are a few tools you need to have installed on your computer before starting with ABP Framework.

IDE/Editor

This book assumes that you are using **Visual Studio 2022** (v10.0 with .NET 6.0 support) or later. If you haven't installed it, the **Community Edition** is freely available on `https://visualstudio.microsoft.com`. However, you can use your favorite **integrated development environment** (**IDE**) or editor, as long as it supports .NET application development with C#.

.NET 6 SDK

If you've installed Visual Studio, you will already have installed the **.NET software development kit (SDK)**. Otherwise, please install .NET 6.0 or later from `https://dotnet.microsoft.com/download`.

Database management system

ABP Framework can work with any data source. However, two main providers are pre-integrated: **Entity Framework Core** (**EF Core**) and **MongoDB**. For EF Core, all **database management systems** (**DBMS**) can be used, such as SQL Server, MySQL, PostgreSQL, Oracle, and so on.

I will use **SQL Server** as the DBMS in this chapter. The startup solution uses **LocalDB**, a simple SQL Server instance for developers installed with Visual Studio. However, you may want to use the full version of SQL Server. In this case, you can download **SQL Server Developer Edition** from `https://www.microsoft.com/sql-server/sql-server-downloads`.

Installing the ABP CLI

Many modern frameworks provide a CLI, and ABP Framework is no exception. **ABP CLI** is a command-line utility to perform some common tasks for ABP applications. It is used to create a new solution with ABP Framework as a fundamental functionality.

Install it using a terminal with the following command:

```
dotnet tool install -g Volo.Abp.Cli
```

If you've already installed it, you can update it to the latest version using the following command:

```
dotnet tool update -g Volo.Abp.Cli
```

We are now ready to create new ABP solutions.

Creating a new solution

ABP Framework provides a pre-built application startup template. There are two ways to create a new solution (project) using this template, which we will explore now.

Downloading the startup solution

You can directly create and download a solution from `https://abp.io/ get-started`. On this page, as shown in the following screenshot, you can easily select the **user interface** (**UI**) framework, database provider, and other available options:

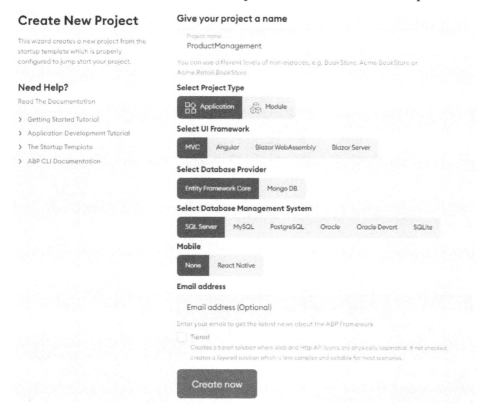

Figure 2.1 – Downloading a new solution

It is worth mentioning the options on this page because they directly affect your solution's architecture, structure, and tooling.

In the **Project name** field is the name of your Visual Studio solution (the .sln file) and the root namespace of your code base.

For **Project type**, there are two options, as follows:

- The **Module** template is used to create reusable application modules.

- The **Application** template is used to build web applications using ABP Framework.

Working with the **Module** template will be covered in *Chapter 15, Working with Modularity*. Here, I chose the **Application** template since I want to create a new web application that we will use in the next chapter.

There are four **UI Framework** options available at the time of writing this book, as follows:

- MVC/Razor Page

- Angular

- Blazor WebAssembly

- Blazor Server

You can select the option that best fits your application requirements and personal or team skills. We will cover the **MVC/Razor Page** and **Blazor** options in *Part 4, User Interface and API Development,* of this book. You can learn more about the Angular UI in ABP's documentation. Here, I select the **MVC/Razor Page** option since we will use it in the next chapter.

There are two database provider options available at the time of writing this book, as follows:

- Entity Framework Core

- MongoDB

If you select the **Entity Framework Core** option, you can use any DBMS supported by EF Core. I've selected EF Core with the **SQLServer** option here.

ABP also provides a mobile startup template based on **React Native**, a popular **single-page application (SPA)** framework provided by Facebook. It provides a good starting point for your mobile application integrated with the same backend if you select it. This book doesn't cover mobile development, so I left that as **None**.

Finally, the **Tiered** option can be checked if you want to separate your UI application from the **HTTP API** physically. In this case, the UI application won't have a direct database connection and perform all operations through the HTTP API. You can deploy the UI and HTTP API applications to separate servers. I haven't checked it to keep it simpler and focus on the ABP features rather than the complexities of distributed systems. However, ABP supports such distributed scenarios as well. You can learn more from ABP's documentation.

When you select the options, ABP creates a fully working, production-ready solution, on top of which you can start to build your application. If you later want to change the options (for example, if you want to use MongoDB instead of EF Core), you should recreate your solution or manually change and configure the NuGet packages. There is no *auto-magic* way of changing these options after creating and customizing your solution.

Downloading your solution from the website makes it easy to see and select the options. However, there is an alternative way for users who like command-line tools.

Using the ABP CLI

Alternatively, you can use the new command in the ABP CLI to create new solutions. Open a command-line terminal and type the following command into an empty directory:

```
abp new ProductManagement
```

ProductManagement is the solution name here. This command creates a new web application using EF Core with SQL Server LocalDB and the MVC/Razor Pages UI because these options are default. If I want to specify all, I can rewrite the same command, like this:

```
abp new ProductManagement -t app -u mvc -d ef -dbms SqlServer
--mobile none
```

If you want to specify the database connection string, you can also pass the --connection-string parameter, as shown in the following example:

```
abp new ProductManagement -t app -u mvc -d ef -dbms SqlServer
--mobile none --connection-string "Server=(LocalDb)\\
MSSQLLocalDB;Database=ProductManagement;Trusted_
Connection=True"
```

The connection string in this example is already the default connection string value and uses LocalDb. See *The connection string* section in this chapter, if you need to change the connection string later.

Please refer to the ABP CLI documentation for all possible options and values: https://docs.abp.io/en/abp/latest/CLI.

> **About the Example Application**
>
> In the next chapter, we will build an example application named ProductManagement. You can use the solution you are currently creating as the starting point for the next chapter.

We now have a well-architected, production-ready solution. The next section shows how to run this solution.

Running the solution

We can use an IDE or code editor to open the solution, create a database, and run the web application. Open the ProductManagement.sln solution in Visual Studio or your favorite IDE. You will see a solution structure like the one depicted here:

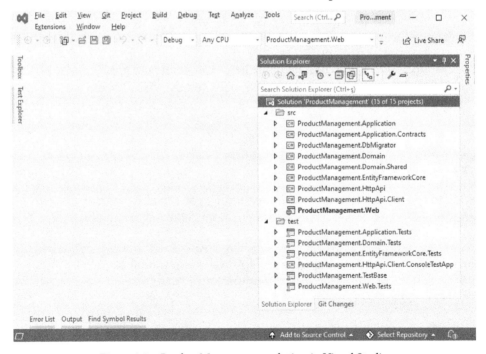

Figure 2.2 – ProductManagement solution in Visual Studio

The solution is layered and contains multiple projects. The test folder has projects to test these layers. Most of these projects are class libraries, while a few of them are executable applications. These are described here:

- `ProductManagement.Web` is the main web application of the solution.
- `ProductManagement.DbMigrator` is used to apply database migrations and seeds the initial data.

The solution uses a database. Before creating a database, you may want to check and change the database connection string.

The connection string

A connection string is used to connect to the database and typically includes the server, database name, and credentials. The connection string is defined in the `appsettings.json` file in the `ProductManagement.Web` and `ProductManagement.DbMigrator` projects, as illustrated in the following code snippet:

```
"ConnectionStrings": {
    "Default": "Server=(LocalDb)\\
MSSQLLocalDB;Database=ProductManagement;Trusted_
Connection=True"
}
```

The default connection string uses `LocalDb`, a lightweight, SQL Server-compatible database for development purposes. It is installed within Visual Studio. If you want to connect to another SQL Server instance, you can change it. If you change it, change it in both places.

This connection string will be used when you create a database in the next section.

Creating a database

The solution uses EF Core code first database migrations. So, we can manage database schema changes with code, using the standard `Add-Migration` and `Update-Database` commands.

`ProductManagement.DbMigrator` is a console application that simplifies creating and migrating a database in development and production. It also seeds the initial data, creating an `admin` role and user to log in to the application.

Right-click the `ProductManagement.DbMigrator` project and select the **Set as Startup Project** command. Then, run the project using *Ctrl + F5* to run it without debugging.

> **About the Initial Migration**
>
> If you are using an IDE other than Visual Studio (for example, JetBrains Rider), you may have problems for the first run since it adds the initial migration and compiles the project. In this case, open a command-line terminal in the directory of the `ProductManagement.DbMigrator` project and execute the `dotnet run` command. For the next time, you can just run it in your IDE as you normally do.

The database is ready, so we can finally run the application to explore the UI.

Running the web application

Set `ProductManagement.Web` as the startup project and run it using *Ctrl + F5* (start without debugging).

> **Tip: Start Without Debugging**
>
> It is strongly suggested to run applications without debugging unless you need to debug them, as this will be much faster.

This will open a landing page where you can delete the content and build your own home page of the application. When you click on the **Login** button, you are redirected to the login page, as illustrated in the following screenshot:

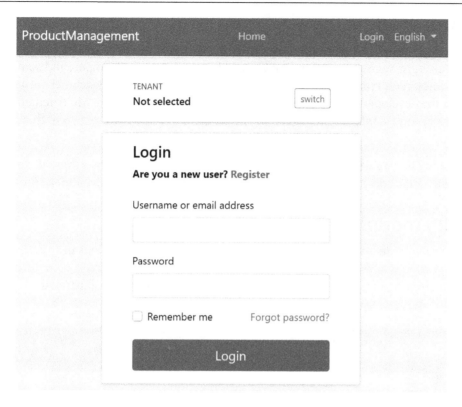

Figure 2.3 – Login page of the application

The default username is admin, and the default password is 1q2w3E*. You can change it after logging in to the application.

ABP is a modular framework, and the startup solution has installed the fundamental modules. Before starting to build your application, it is good to explore the pre-built module functionalities.

Exploring the pre-built modules

This section will explore the fundamental modules pre-installed in the startup solution: **Account**, **Identity**, and **Tenant Management**.

The source code of these modules is not included in the download solution by default, but they are freely available on GitHub. They are used as **NuGet** packages and easily upgraded when a new ABP version is published. They are designed as highly customizable, without touching their code. However, if you need, you can include their source code in your solution to freely change them based on your unique requirements.

Let's start with the Account module, which provides user authentication features.

Account module

The login page shown in *Figure 2.3* comes from the **Account** module. This module implements login, registering, a forgot password functionality, social logins, and some other common requirements. It also shows a tenant selection area to switch between tenants in the development environment for a multi-tenant application. Multi-tenancy will be covered in *Chapter 16, Implementing Multi-Tenancy*, so we will return to this screen again.

When you log in, you will see an **Administration** menu item with a few submenu items. These menu items come with ABP's pre-built **Identity** and **Tenant Management** modules.

Identity module

The **Identity** module is used to manage users, roles, and their permissions in your application. It adds an **Identity management** menu item under the **Administration** menu, with **Roles** and **Users** as its submenu items, as illustrated in the following screenshot:

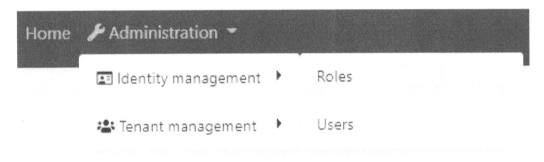

Figure 2.4 – Identity management menu

If you click on the **Roles** menu item, the role management page is opened, as illustrated in the following screenshot:

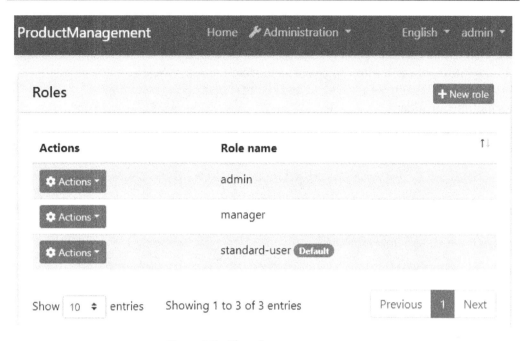

Figure 2.5 – The role management page

On this page, you can manage roles and their permissions in your application. In ABP, a role is a group of privileges. Roles are assigned to users to authorize them. The **Default** badge in *Figure 2.5* indicates the default role. Default roles are automatically assigned to new users when they are registered with the system. We will return to the **Roles** page in the *Working with authorization and permission systems* section of *Chapter 7, Exploring Cross-Cutting Concerns*.

The **Users** page, on the other hand, is used to manage the users in your application. A user can have zero or more roles.

Roles and users are pretty standard almost in all business applications, while the **Tenant Management** page module is only used in multi-tenancy systems.

Tenant Management module

The **Tenant Management** module is where you create and manage your tenants in a multi-tenant system. In a multi-tenant application, a tenant is a customer that has its own data—including roles, users, and permissions—isolated from the other tenants. It is an efficient and common way of building **Software as a Service (SaaS)** solutions. If your application is not multi-tenant, you can just remove this module from your solution.

The Tenant Management module and multi-tenancy will be covered in *Chapter 16, Implementing Multi-Tenancy*.

Summary

In this chapter, we've installed some required tools to prepare our development environment. Then, we saw how to create a new solution using the direct download and CLI options. Finally, we configured and run the application to explore the pre-built functionalities.

In the next chapter, we will learn how to add our own functionalities to this startup solution by understanding the solution structure.

3

Step-By-Step Application Development

This chapter introduces the fundamentals of ABP Framework by building an example application. The example application is used to manage products on a typical **CRUD** page (note that a CRUD page is used to **Create**, **Read** (view), **Update**, and **Delete** entities).

The example presented in this chapter is more advanced than a simple CRUD page. It implements many aspects of application development with production quality. By the end of this chapter, you will understand the basics, and you will be ready to start development with ABP Framework.

I will proceed, step by step, in the order of building a real-world project. This chapter consists of the following topics; each represents a step in this process:

- Creating the solution
- Defining the domain objects
- **Entity Framework (EF)** Core and database mappings
- Listing the product data

- Creating products
- Editing products
- Deleting products

User Interface (UI) and Database Preference

I prefer **Razor Pages (MVC)** as the UI framework and **EF Core** as the database provider. We will cover other UI frameworks and database providers in separate chapters.

Technical requirements

We will be building an application, so you need to have .NET runtime, ABP CLI, and an IDE/editor installed to build ASP.NET Core projects.

Please refer to *Chapter 2, Getting Started with ABP Framework*, to learn how to prepare your development environment, as well as create and run the solution.

You can download the source code of the final application from the GitHub repository at `https://github.com/PacktPublishing/Mastering-ABP-Framework`.

Creating the solution

The first step is to create a solution for the product management application. If you've created the *ProductManagement* solution in *Chapter 2, Getting Started with ABP Framework*, you can use it. Otherwise, create an empty folder in your computer, open a command-line terminal in this folder, and run the following **ABP CLI** command to create a new web application:

```
abp new ProductManagement -t app
```

Open the solution in your favorite IDE, create the database, and run the web project. If you have problems with running the solution, please refer to the previous chapter.

Now we have a running solution. We can start the development by defining the domain objects of the solution.

Defining the domain objects

In this section, you will learn how to define entities with ABP Framework. The domain is simple for this application. We have **Product** and **Category** entities and a **ProductStockState** enum, as shown in *Figure 3.1*:

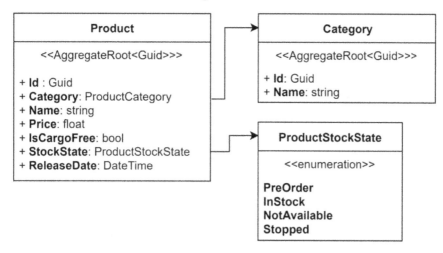

Figure 3.1 – An example product management domain

Entities are defined in the *Domain Layer* of the solution, and the domain layer is split into two projects within the solution:

- **ProductManagement.Domain** is used to define your entities, value objects, domain services, repository interfaces, and other core domain-related classes.

- **ProductManagement.Domain.Shared** is used to define some primitive shared types. The types defined in this project are available to all other layers. Typically, we define enums and some constants here.

So, we can start by creating the Category and Product entities and the ProductStockState enum.

Category

The Category entity is used to categorize the products. Create a *Categories* folder inside the *ProductManagement.Domain* project and a Category class inside it:

```
using System;
using Volo.Abp.Domain.Entities.Auditing;
namespace ProductManagement.Categories
{
```

```
public class Category : AuditedAggregateRoot<Guid>
{
    public string Name { get; set; }
}
}
```

`Category` is a class is derived from `AuditedAggregateRoot<Guid>`. Here, `Guid` is the primary key (`Id`) type of the entity. You can use any type of primary key (such as `int`, `long`, or `string`) as long as your database management system supports it.

`AggregateRoot` is a special type of entity that is used to create the root entity type of an aggregate. An aggregate is a **Domain-Driven Design** (**DDD**) concept that we will discuss in greater detail in the upcoming chapters. For now, consider that we inherit the main entities from this class.

The `AuditedAggregateRoot` class adds some more properties to the `AggregateRoot` class: `CreationTime` as `DateTime`, `CreatorId` as `Guid`, `LastModificationTime` as `DateTime`, and `LastModifierId` as `Guid`.

ABP automatically sets these properties. For example, when you insert an entity into the database, `CreationTime` is set to the current time, and `CreatorId` is automatically set to the `Id` property of the current user.

The audit logging system and the base `Audited` classes will be covered in *Chapter 8, Using the Features and Services of ABP*.

About Rich Domain Models

In this chapter, I keep the entities simple, with public getters and setters. If you want to create rich domain models and apply DDD principles and other best practices, we will discuss them in upcoming chapters.

ProductStockState

`ProductStockState` is a simple enum to set and track the availability of the product in stock.

Create a *Products* folder inside the *ProductManagement.Domain.Shared* project and a `ProductStockState` enum inside it:

```
namespace ProductManagement.Products
{
    public enum ProductStockState : byte
```

```
    {
        PreOrder,
        InStock,
        NotAvailable,
        Stopped
    }
}
```

We define this enum in the `ProductManagement.Domain.Shared` project since we will reuse it in the **Data Transfer Objects (DTOs)** and the UI layer.

Product

The `Product` class represents a real product. I intentionally added different types of properties to show their usages. Create a *Products* folder inside the *ProductManagement. Domain* project and a `Product` class inside it:

```
using System;
using Volo.Abp.Domain.Entities.Auditing;
using ProductManagement.Categories;


namespace ProductManagement.Products
{
    public class Product : FullAuditedAggregateRoot<Guid>
    {
        public Category Category { get; set; }
        public Guid CategoryId { get; set; }
        public string Name { get; set; }
        public float Price { get; set; }
        public bool IsFreeCargo { get; set; }
        public DateTime ReleaseDate { get; set; }
        public ProductStockState StockState { get; set; }
    }
}
```

This time, I inherited from `FullAuditedAggregateRoot`, which adds `IsDeleted` as `bool`, `DeletionTime` as `DateTime`, and `DeleterId` as `Guid` properties in addition to the `AuditedAggregateRoot` class used for the `Category` class.

`FullAuditedAggregateRoot` implements the `ISoftDelete` interface, which makes the entity **Soft-Delete**. That means it is never deleted from the database but just *marked as deleted*. ABP automatically handles all the Soft-Delete logic. You delete the entity as you normally do, but it is not actually deleted. The next time you query, deleted entities are automatically filtered, and you don't get them in the query result unless you intentionally request them. We will return to that feature in the *Using the data filtering system* section of *Chapter 8, Using the Features and Services of ABP.*

About the Navigation Properties

In this example, `Product.Category` is a navigation property for the `Category` entity. If you use MongoDB or want to implement DDD truly, you should not add navigation properties to other aggregates. However, for relational databases, it works perfectly and provides flexibility to our code. We will discuss alternative approaches in *Chapter 10, DDD – The Domain Layer.*

The new files in the solution should look like *Figure 3.2*:

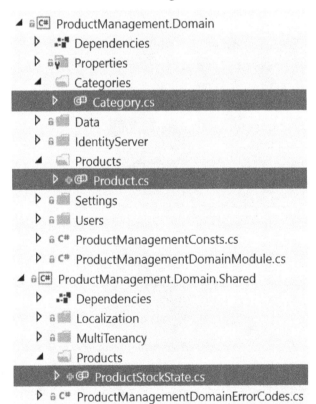

Figure 3.2 – Adding domain objects to the solution

We've created the domain objects. In addition, we will create a few `const` values to be used later in the application.

Constants

We need to define constant values for the properties of the entities. We will then use them in the input validation and database mapping phase.

First, create a *Categories* folder inside the *ProductManagement.Domain.Shared* project and add a `CategoryConsts` class inside it:

```
namespace ProductManagement.Categories
{
    public static class CategoryConsts
    {
        public const int MaxNameLength = 128;
    }
}
```

Here, the `MaxNameLength` value will be used to implement the constraint for the `Name` property of a `Category` instance.

Then, create a `ProductConsts` class inside the *Products* folder of the *ProductManagement.Domain.Shared* project:

```
namespace ProductManagement.Products
{
    public static class ProductConsts
    {
        public const int MaxNameLength = 128;
    }
}
```

The `MaxNameLength` value will be used to implement the constraint for a `Product` instance's `Name` property.

The *ProductManagement.Domain.Shared* project should look similar to *Figure 3.3*:

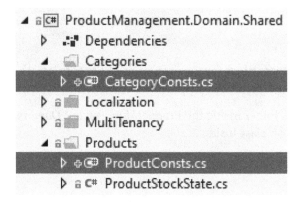

Figure 3.3 – Adding constant classes

Now that the domain layer has been completed, we can now configure the database mappings for EF Core.

EF Core and database mappings

We are using **EF Core** in this application. EF Core is an **Object-Relational Mapping (ORM)** provider provided by Microsoft. ORMs provide abstractions to make you feel like you are working with objects in your application code rather than the database tables. We will cover ABP's EF Core integration in *Chapter 6, Working with the Data Access Infrastructure*. However, for now, let's focus on how we can use it practically.

First, we will add entities to the DbContext class and define the mappings between entities and database tables. Then, we will use EF Core's **Code First Migration** approach to build the necessary code that creates the database tables. Following this, we will look at ABP's **Data Seeding** system to insert some initial data into the database. Finally, we will apply the migrations and seed data to the database to prepare it for the application.

First, let's start by defining the DbSet properties for the entities.

Adding entities to the DbContext class

EF's DbContext class is the main class that is used to define mappings between entities and database tables. Additionally, it is used to access the database and perform database operations for the related entities.

Open the `ProductManagementDbContext` class in the *ProductManagement.EntityFrameworkCore* project, and add the following `DbSet` properties inside it (you will need to import the namespaces of the `Product` and `Category` objects):

```
public DbSet<Product> Products { get; set; }
public DbSet<Category> Categories { get; set; }
```

Adding a `DbSet` property for an entity relates the entity with the `DbContext` class. Then, we can use that `DbContext` class to perform database operations for the entity. EF Core can make most of the mapping using conventions based on the property names and types. If you want to customize the default mapping configuration or perform additional configurations, you have two options: **Data Annotations** (attributes) and **Fluent API**.

In the data annotation approach, you add attributes, such as `[Required]` and `[StringLength]`, to your entity properties. It is very practical and easy to use. It also makes it easier to understand when you read the source code of your entity.

One problem with the data annotation attributes is that they are limited (compared to the Fluent API) and make your domain layer dependant on the EF Core NuGet package when you need to use EF Core's custom attributes, such as `[Index]` and `[Owned]`. If that's not a problem for you, you can use the data annotation attributes and combine them with the Fluent API where they are not sufficient.

In this chapter, I will prefer the Fluent API approach, which keeps the entity cleaner and places all the ORM logic inside the infrastructure layer.

Mapping entities to the database tables

The `ProductManagementDbContext` class (in the *ProductManagement.EntityFrameworkCore* project) contains an `OnModelCreating` method to configure mappings of the entities to the database tables. When you first create your solution, this method looks like the following:

```
protected override void OnModelCreating(ModelBuilder builder)
{
    base.OnModelCreating(builder);

    builder.ConfigurePermissionManagement();
    builder.ConfigureSettingManagement();
    builder.ConfigureIdentity();
    ...configuration of the other modules
```

```
    /* Configure your own tables/entities here */
}
```

Add the following code after the preceding comment:

```
builder.Entity<Category>(b =>
{
        b.ToTable("Categories");
        b.Property(x => x.Name)
            .HasMaxLength(CategoryConsts.MaxNameLength)
            .IsRequired();
        b.HasIndex(x => x.Name);
});

builder.Entity<Product>(b =>
{
        b.ToTable("Products");
        b.Property(x => x.Name)
            .HasMaxLength(ProductConsts.MaxNameLength)
            .IsRequired();
        b.HasOne(x => x.Category)
            .WithMany()
            .HasForeignKey(x => x.CategoryId)
            .OnDelete(DeleteBehavior.Restrict)
            .IsRequired();
    b.HasIndex(x => x.Name).IsUnique();
});
```

This code part defines the `Category` and `Product` mapping configurations.

> **About Namespaces**
>
> You might need to add `using` statements for the namespaces of the `Product` class, the `Category` class, and any other classes used in the code. If you have trouble, you can always refer to the source code in the GitHub repository that I've shared in the *Technical requirements* section of this chapter.

The Category entity is mapped to the *Categories* database table. We use CategoryConsts.MaxNameLength that was defined before to set the maximum length of the Name field in the database. The Name field is also a *required* property. Finally, we define a *unique* database index for the Name property because it helps search categories by the Name field.

The Product mapping is similar to the Category mapping. Additionally, it defines a relationship between the Category entity and the Product entity; a Product entity belongs to a Category entity, while a Category entity can have many related Product entities.

> **EF Core Fluent Mapping**
>
> You can refer to the EF Core documentation to learn about all the details and other options for the Fluent Mapping API.

The mapping configuration is complete. It is time to create a database migration to update the database schema for the newly added entities.

The Add-Migration command

When you create a new entity or make changes to an existing entity, you should also create or alter the related table in the database. EF Core's **Code First Migration** system is a perfect way to keep the database schema aligned with the application code. Typically, you generate migrations and apply them to the database. A migration is an incremental schema change for the database. When you update the database, all the migrations are applied since the last update, and the database becomes aligned with the application code.

There are two ways to generate a new migration.

With Visual Studio

If you are using Visual Studio, open **Package Manager Console (PMC)** from the **View | Other Windows | Package Manager Console** menu:

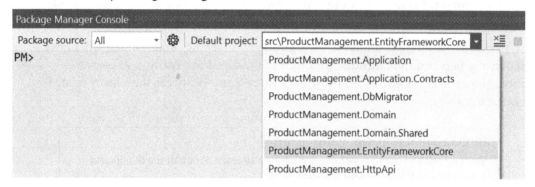

Figure 3.4 – Package Manager Console

Select the *ProductManagement.EntityFrameworkCore* project as the **Default project** type. Ensure that the *ProductManagement.Web* project is selected as the startup project. You can right-click on the *ProductManagement.Web* project and click on the **Set as Startup Project** action.

Now, you can type the following command into the PMC to add a new migration class:

```
Add-Migration "Added_Categories_And_Products"
```

The output of this command should be similar to *Figure 3.5*:

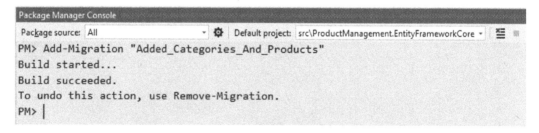

Figure 3.5 – Output of the Add-Migration command

If you get an error such as *No DbContext was found in assembly...*, be sure that you've set the **Default project** type to the *ProductManagement.EntityFrameworkCore* project.

If everything goes well, a new migration class should be added inside the *Migrations* folder of the *ProductManagement.EntityFrameworkCore* project.

In the command line

If you are not using Visual Studio, you can use the EF Core command-line tools. If you haven't installed it yet, execute the following command in a command-line terminal:

```
dotnet tool install --global dotnet-ef
```

Now, open a command-line terminal in the root directory of the *ProductManagement. EntityFrameworkCore* project, and type in the following command:

```
dotnet ef migrations add "Added_Categories_And_Products"
```

A new migration class should be added inside the *Migrations* folder of the *ProductManagement.EntityFrameworkCore* project.

Before applying the newly created migration to the database, I want to mention the data seeding feature of ABP Framework.

Seeding data

The data seeding system is used to add some initial data when you migrate the database. For example, the identity module creates an admin user in the database with all permissions granted to log in to the application.

While data seeding is not essential in our scenario, I want to add some example categories and products to the database to make it easier to develop and test the application.

> **About the EF Core Data Seeding**
>
> This section uses ABP's data seed system, while EF Core has its own data seeding feature. The ABP data seed system allows you to inject runtime services and implement advanced logic in your data seed code, and it is suitable for development, test, and production environments. However, for simpler development and test scenarios, you can use EF Core's data seeding system. Please check the official documentation at https://docs.microsoft. com/en-us/ef/core/modeling/data-seeding.

Create a ProductManagementDataSeedContributor class in the *Data* folder of the *ProductManagement.Domain* project:

```
using ProductManagement.Categories;
using ProductManagement.Products;
using System;
using System.Threading.Tasks;
```

```
using Volo.Abp.Data;
using Volo.Abp.DependencyInjection;
using Volo.Abp.Domain.Repositories;
namespace ProductManagement.Data
{
    public class ProductManagementDataSeedContributor :
            IDataSeedContributor, ITransientDependency
    {
        private readonly IRepository<Category,
Guid>_categoryRepository;
        private readonly IRepository<Product,
Guid>_productRepository;

        public ProductManagementDataSeedContributor(
            IRepository<Category, Guid> categoryRepository,
            IRepository<Product, Guid> productRepository)
        {
            _categoryRepository = categoryRepository;
            _productRepository = productRepository;
        }

        public async Task SeedAsync(DataSeedContext
context)
        {
            /***** TODO: Seed initial data here *****/
        }
    }
}
```

This class implements the IDataSeedContributor interface. ABP automatically discovers and calls its SeedAsync method when you want to seed the database. You can implement constructor injection and use any service in your class (such as the repositories in this example).

Then, write the following code inside the SeedAsync method:

```
if (await _categoryRepository.CountAsync() > 0)
{
    return;
```

```
}

var monitors = new Category { Name = "Monitors" };
var printers = new Category { Name = "Printers" };

await _categoryRepository
    .InsertManyAsync(new[] { monitors, printers });

var monitor1 = new Product
{
    Category = monitors,
    Name = "XP VH240a 23.8-Inch Full HD 1080p IPS LED
Monitor",
    Price = 163,
    ReleaseDate = new DateTime(2019, 05, 24),
    StockState = ProductStockState.InStock
};

var monitor2 = new Product
{
    Category = monitors,
    Name = "Clips 328E1CA 32-Inch Curved Monitor, 4K UHD",
    Price = 349,
    IsFreeCargo = true,
    ReleaseDate = new DateTime(2022, 02, 01),
    StockState = ProductStockState.PreOrder
};

var printer1 = new Product
{
    Category = monitors,
    Name = "Acme Monochrome Laser Printer, Compact All-In
One",
    Price = 199,
    ReleaseDate = new DateTime(2020, 11, 16),
    StockState = ProductStockState.NotAvailable
};
```

```
await _productRepository
    .InsertManyAsync(new[] { monitor1, monitor2, printer1 });
```

We've created two categories with three products and inserted them into the database. This class is executed whenever you run the *DbMigrator* application (please refer to the following section). Additionally, we checked `if (await _categoryRepository. CountAsync() > 0)` to prevent us from inserting the same data in every run.

We are now ready to migrate the database, which will update the database schema and seed the initial data.

Migrating the database

The ABP application startup template includes a *DbMigrator* console application that is pretty useful in development and production environments. When you run it, all pending migrations are applied in the database, and the data seeder classes are executed. It supports multi-tenant, multi-database scenarios, which is not possible if you use the standard `Update-Database` command. This application can be deployed and executed in the production environment, typically, as a stage of your **Continuous Deployment** (**CD**) pipeline. Separating the migration from the main application is a good approach, as the main application doesn't require permission to alter the database schema in such cases. Also, you can get rid of any concurrency issues you might have if you apply migrations in your main application and run multiple instances of the application.

Run the *ProductManagement.DbMigrator* application to migrate the database (that is, set it as the startup project, and hit *Ctrl + F5*). Once the application exits, you can check the database to see that the *Categories* and *Products* tables have the initial data inserted (if you are using Visual Studio, you can use **SQL Server Object Explorer** to connect to **LocalDB** and explore the databases).

The EF Core configuration is complete, and the database is ready for development. We will continue by showing the product data on the UI.

Listing the product data

I prefer to develop the application functionality feature by feature. This section will explain how to show a list of the products in a data table on the UI.

We will begin by defining a **DTO**, `ProductDto`, for the `Product` entity. Then, we will create an application service method that returns a list of products to the presentation layer. Additionally, we will learn how to map the `Product` entity to `ProductDto` automatically.

Before creating the UI, I will show you how to write an **Automated Test** for the application service. In this way, we will be sure that the application service is working properly before starting the UI development.

Throughout the development, we will explore some benefits of ABP Framework, such as the automatic API controller and dynamic JavaScript Proxy systems.

Finally, we will create a new page, add a data table inside it, get a list of the products from the server, and show it on the UI.

In the next section, we will begin by creating a `ProductDto` class.

The ProductDto class

DTOs are used to transfer data between the application and presentation layers. It is best practice to return DTOs to the presentation (UI) layer instead of the entities. DTOs allow you to expose data in a controlled way and abstract your entities from the presentation layer. Directly exposing entities to the presentation layer might cause serialization and security problems, too. We will discuss the benefits of using DTOs in *Chapter 11, DDD – The Application Layer.*

DTOs are defined in the *Application.Contracts* project to make them available within the UI layer. So, we start by creating a `ProductDto` class inside the *Products* folder of the *ProductManagement.Application.Contracts* project:

```
using System;
using Volo.Abp.Application.Dtos;
namespace ProductManagement.Products
{
    public class ProductDto : AuditedEntityDto<Guid>
    {
        public Guid CategoryId { get; set; }
        public string CategoryName { get; set; }
        public string Name { get; set; }
        public float Price { get; set; }
        public bool IsFreeCargo { get; set; }
        public DateTime ReleaseDate { get; set; }
```

```
        public ProductStockState StockState { get; set; }
    }
}
```

The `ProductDto` class is a similar class to the `Product` entity with the following differences:

- It is derived from `AuditedEntityDto<Guid>`, which defines the `Id`, `CreationTime`, `CreatorId`, `LastModificationTime`, and `LastModifierId` properties (we don't need to delete auditing properties, such as `DeletionTime`, since the deleted entities are not read from the database).

- Instead of adding a navigation property to the `Category` entity, we used a `string` `CategoryName` property, which is enough to show on the UI.

We will use the `ProductDto` class to return a list of products from the `IProductAppService` interface.

IProductAppService

Application Services implement the use cases of an application. The UI uses them to perform business logic on user interactions. Typically, an application service method gets and returns DTOs.

Application Services versus API Controllers

You could compare application services with API controllers in an ASP. NET Core MVC application. While they have similarities for some use cases, application services are plain classes that better fit into DDD. They don't depend on a particular UI technology. In addition, ABP can automatically expose your application services as HTTP APIs, as we will discover in the *Auto API Controllers and the Swagger UI* section of this chapter.

We define interfaces for application services in the *Application.Contracts* project of the solution. Create an `IProductAppService` interface inside the *Products* folder of the *ProductManagement.Application.Contracts* project:

```
using System.Threading.Tasks;
using Volo.Abp.Application.Dtos;
using Volo.Abp.Application.Services;
namespace ProductManagement.Products
{
```

```
    public interface IProductAppService :
IApplicationService
    {
        Task<PagedResultDto<ProductDto>>
            GetListAsync(PagedAndSortedResultRequestDto
input);
    }
}
```

You can see some predefined ABP types in the preceding code block:

- `IProductAppService` is derived from the `IApplicationService` interface. In this way, ABP can recognize the application services.

- The `GetListAsync` method gets `PagedAndSortedResultRequestDto`, which is a standard DTO class of ABP Framework that defines the `MaxResultCount` (int), `SkipCount` (int), and `Sorting` (string) properties.

- The `GetListAsync` method returns `PagedResultDto<ProductDto>`, which contains a `TotalCount` (long) property and an `Items` collection of `ProductDto` objects. That is a convenient way of returning paged results with ABP Framework.

You could use your own DTOs instead of these predefined DTO types. However, they are pretty helpful when you want to standardize some common patterns and use the same naming everywhere.

> **Asynchronous Methods**
>
> It is a best practice to define all the application service methods as asynchronous. If you define synchronous application service methods, in some cases, certain ABP features (such as Unit of Work) might not work as expected.

Now, we can implement the `IProductAppService` interface to perform the use case.

ProductAppService

Create a `ProductAppService` class inside the *Products* folder of the *ProductManagement.Application* project:

```
using System;
using System.Collections.Generic;
using System.Linq;
using System.Linq.Dynamic.Core;
```

```
using System.Threading.Tasks;
using Volo.Abp.Application.Dtos;
using Volo.Abp.Domain.Repositories;
namespace ProductManagement.Products
{
    public class ProductAppService :
        ProductManagementAppService, IProductAppService
    {
        private readonly IRepository<Product, Guid>
_productRepository;

        public ProductAppService(
            IRepository<Product, Guid> productRepository)
        {
            _productRepository = productRepository;
        }

        public async Task<PagedResultDto<ProductDto>>
GetListAsync(
            PagedAndSortedResultRequestDto input)
        {
            /* TODO: Implementation */
        }
    }
}
```

The `ProductAppService` class is derived from `ProductManagementAppService`, which was defined in the startup template, and can be used as the base class for your application services. It implements the `IProductAppService` interface that was previously defined. It injects the `IRepository<Product, Guid>` service. This is called a **Default Repository**. A repository is a collection-like interface that allows you to perform operations on the database. ABP automatically provides default repository implementations for all aggregate root entities.

We can implement the `GetListAsync` method, as shown in the following code block:

```
public async Task<PagedResultDto<ProductDto>> GetListAsync(
    PagedAndSortedResultRequestDto input)
{
```

```
    var queryable = await _productRepository
        .WithDetailsAsync(x => x.Category);

    queryable = queryable
        .Skip(input.SkipCount)
        .Take(input.MaxResultCount)
        .OrderBy(input.Sorting ?? nameof(Product.Name));

    var products = await
AsyncExecuter.ToListAsync(queryable);
    var count = await _productRepository.GetCountAsync();

    return new PagedResultDto<ProductDto>(
        count,
        ObjectMapper.Map<List<Product>, List<ProductDto>>
(products)
    );
}
```

Here, `_productRepository.WithDetailsAsync` returns an
`IQueryable<Product>` object by including the categories (the `WithDetailsAsync`
method is similar to EF Core's `Include` extension method, which loads the related data
into the query). We can use the standard **Language-Integrated Query** (**LINQ**) extension
methods such as `Skip`, `Take`, and `OrderBy` on the queryable object.

The `AsyncExecuter` service (which is pre-injected in the base class) is used to execute
the `IQueryable` object to perform a database query asynchronously. This makes it
possible to use the async LINQ extension methods without depending on the EF Core
package in the application layer.

Finally, we are using the `ObjectMapper` service (pre-injected in the base class) to map a
list of `Product` (entity) objects to a `ProductDto` (DTO) object list. In the next section,
we will explain how the object mapping is configured.

> **Repositories and Async Query Execution**
>
> We will explore `IRepository` and `AsyncExecuter` in greater detail in
> *Chapter 6, Working with the Data Access Infrastructure*.

Object to object mapping

`ObjectMapper` (the `IObjectMapper` service) automates type conversions and uses the **AutoMapper** library by default. It requires you to define the mapping before using it. The startup template contains a profile class that you can create the mappings inside.

Open the `ProductManagementApplicationAutoMapperProfile` class in the *ProductManagement.Application* project, and change it to the following:

```
using AutoMapper;
using ProductManagement.Products;
namespace ProductManagement
{
    public class ProductManagementApplicationAutoMapperProfile
        : Profile
    {
        public ProductManagementApplicationAutoMapperProfile()
        {
            CreateMap<Product, ProductDto>();
        }
    }
}
```

Here, `CreateMap` defines the mapping. Then, you can automatically convert `Product` objects into `ProductDto` objects where you need them.

One of the interesting AutoMapper features is **Flattening**. This involves taking a complex object model and flattening it into a simpler model. In this example, the `Product` class has a `Category` property, and the `Category` class has a `Name` property. So, if you want to access the category name of a product, you should use the `Product.Category.Name` expression. However, `ProductDto` has a direct `CategoryName` property that can be accessed using the `ProductDto.CategoryName` expression. AutoMapper automatically maps these expressions by flattening `Category.Name` into `CategoryName`.

The application layer is complete. Before starting the UI, I want to show you how to write automated tests for the application layer.

Testing the ProductAppService class

The startup template comes with the test infrastructure properly configured using the **xUnit**, **Shouldly**, and **NSubstitute** libraries. It uses the *SQLite in-memory* database to mock the database. A separate database is created for each test. It is seeded and destroyed at the end of the test. In this way, tests do not affect each other, and your real database remains untouched.

Chapter 17, Building Automated Tests, will explore all the details of testing. However, here, I want to show you how you can easily write some automated test code for the GetListAsync method of the ProductAppService class. It is good to write the test code for the application services before using them on the UI.

Create a *Products* folder in the *ProductManagement.Application.Tests* project, and create a ProductAppService_Tests class inside it:

```
using Shouldly;
using System.Threading.Tasks;
using Volo.Abp.Application.Dtos;
using Xunit;
namespace ProductManagement.Products
{
    public class ProductAppService_Tests
        : ProductManagementApplicationTestBase
    {
        private readonly IProductAppService
_productAppService;

        public ProductAppService_Tests()
        {
            _productAppService =
                GetRequiredService<IProductAppService>();
        }


        /* TODO: Test methods */
    }
}
```

This class inherits from the `ProductManagementApplicationTestBase` class (which is included in your solution) that integrates ABP Framework and other infrastructure libraries and makes it possible to write our tests. Instead of constructor injection (which is not possible in tests), we use the `GetRequiredService` method to resolve dependencies in the test code.

Now, we can write the first test method. Add the following method inside the `ProductAppService_Tests` class:

```
[Fact]
public async Task Should_Get_Product_List()
{
    //Act
    var output = await _productAppService.GetListAsync(
        new PagedAndSortedResultRequestDto()
    );

    //Assert
    output.TotalCount.ShouldBe(3);
    output.Items.ShouldContain(
        x => x.Name.Contains("Acme Monochrome Laser
Printer")
    );
}
```

This method calls the `GetListAsync` method and checks whether the result is correct. If you open the **Test Explorer** window (under the **View | Test Explorer** menu in Visual Studio), you can see the test method that we've added. **Test Explorer** is used for showing and running the tests in the solution:

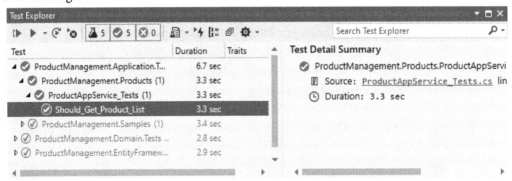

Figure 3.6 – The Test Explorer window

Run the test to check whether it is working as expected. If the `GetListAsync` method works properly, you will see a green icon on the left-hand side of the test method name, as shown in *Figure 3.6*. Unit and integration tests will be covered in *Chapter 17, Building Automated Tests*.

Auto API Controllers and the Swagger UI

Swagger is a popular tool in which to explore and test HTTP APIs. It comes preinstalled with the startup template.

Run the *ProductManagement.Web* project to start the web application (set it as the startup project if not done so already, and hit *Ctrl + F5*). Once the application starts, enter the / `swagger` URL, as shown in *Figure 3.7*:

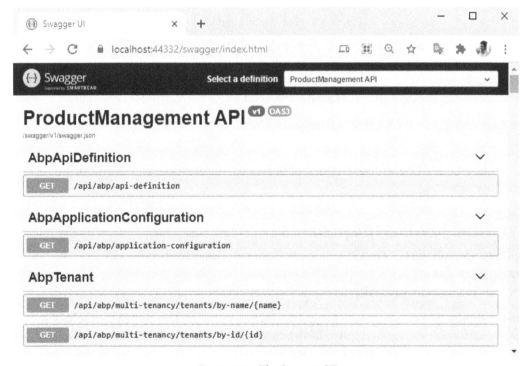

Figure 3.7 – The Swagger UI

You will see a lot of API endpoints coming from the modules installed in the application. If you scroll down, you will see a **Product** endpoint, too. You can test it to get the list of products:

Figure 3.8 – The Product endpoint

We haven't created a *ProductController* endpoint yet. So, how is this endpoint available here? This is known as the **Auto API Controller** feature of ABP Framework. It automatically exposes your application services as HTTP APIs based on naming conventions and configurations. Typically, we don't write the controllers manually.

The Auto API Controller feature will be covered in *Chapter 14, Building HTTP APIs and Real-Time Services*, in detail.

So, we have the HTTP API to get the list of products. The next step is to consume this API from the client code.

Dynamic JavaScript proxies

Typically, you call the HTTP API endpoints from your JavaScript code. ABP dynamically creates client-side proxies for all HTTP APIs. Then, you can use these dynamic JavaScript functions to consume your APIs from the client application.

Run the *ProductManagement.Web* project again, and open the **Developer Console** of the browser while you are on the application's landing page. The developer console is available in any modern browser and is typically opened using the *F12* shortcut key (on Windows). It is used to explore, trace, and debug the application by developers.

Open the **Console** tab, and type in the following JavaScript code:

```
productManagement.products.product.getList({}).
then(function(result) {
    console.log(result);
});
```

Once you execute this code, a request is made to the server, and the returning result is logged in the **Console** tab, as shown in *Figure 3.9*:

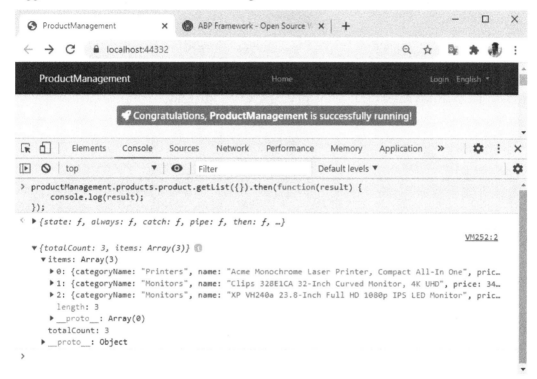

Figure 3.9 – Using the dynamic JavaScript proxies

We can see that the product list is logged in the **Console** tab. That means we can easily consume server-side APIs from JavaScript code without having to deal with the low-level details.

If you are wondering where that `getList` JavaScript is defined, you can check the `/Abp/ServiceProxyScript` endpoint in your application to see the JavaScript proxy functions dynamically created by ABP Framework.

In the next section, we will create a **Razor Page** to show the table of products on the UI.

Creating a products page

Razor Pages is the recommended way of creating a UI in the ASP.NET Core MVC framework.

First, create a *Products* folder under the *Pages* folder of the *ProductManagement.Web* project. Then, add a new, empty razor page by right-clicking on the *Products* folder and then selecting **Add | Razor Page**. Select the **Razor Page - Empty** option. Name it `Index.cshtml`. *Figure 3.10* shows the location of the page that we've added:

Figure 3.10 – Creating a Razor Page

Edit the `Index.cshtml` content, as shown in the following code block:

```
@page
@using ProductManagement.Web.Pages.Products
@model IndexModel


<h1>Products Page</h1>
```

Here, I've just placed an `h1` element as the page header. When we create a page, typically, we want to add an item to the main menu to open this page.

Adding a new menu item

ABP provides a dynamic and modular menu system. Every module can add items to the main menu.

Open the `ProductManagementMenuContributor` class in the *Menus* folder of the *ProductManagement.Web* project, and add the following code at the end of the `ConfigureMainMenuAsync` method:

```
context.Menu.AddItem(
    new ApplicationMenuItem(
        "ProductManagement",
        l["Menu:ProductManagement"],
        icon: "fas fa-shopping-cart"
            ).AddItem(
        new ApplicationMenuItem(
            "ProductManagement.Products",
            l["Menu:Products"],
            url: "/Products"
        )
    )
);
```

This code adds a *Product Management* main menu item with the *Products* menu item inside it. It uses localization keys (with the `l["..."]` syntax) that we should define. Open the `en.json` file in the *Localization/ProductManagement* folder of the *ProductManagement.Domain.Shared* project, and add the following entries to the end of the `texts` section:

```
"Menu:ProductManagement": "Product Management",
"Menu:Products": "Products"
```

Localization keys are arbitrary, which means you can use any string value as the localization key. I prefer to use the `Menu:` prefix for the localization keys of menu items, such as `Menu:Products`, in this example. We will return to the topic of localization in *Chapter 8, Using the Features and Services of ABP*.

Now, you can rerun the application and open the *Products* page using the new *Product Management* menu item, as shown in *Figure 3.11*:

Figure 3.11 – The Products page

So, we've created a page and can open the page using the menu element. We are ready to create a data table to show the list of products.

Creating the products data table

We will create a data table to show the list of products with paging and sorting. The ABP startup template comes with the **Datatables.net** JavaScript library preinstalled and configured. It is a flexible and feature-rich library to show tabular data.

Open the `Index.cshtml` page (in the *Pages/Products* folder), and change its contents to the following:

```
@page
@using ProductManagement.Web.Pages.Products
@using Microsoft.Extensions.Localization
@using ProductManagement.Localization
@model IndexModel
@inject IStringLocalizer<ProductManagementResource> L
@section scripts
{
    <abp-script src="/Pages/Products/Index.cshtml.js" />
}
<abp-card>
    <abp-card-header>
        <h2>@L["Menu:Products"]</h2>
    </abp-card-header>
    <abp-card-body>
        <abp-table id="ProductsTable" striped-rows="true" />
```

```
        </abp-card-body>
    </abp-card>
```

Here, `abp-script` is an ABP tag helper for adding script files to the page with automatic bundling, minification, and versioning support. `abp-card` is another tag helper to render a card component in a type-safe and easy way (it renders a Bootstrap card).

We could use the standard HTML tags. However, ABP tag helpers dramatically simplify UI creation in MVC/Razor Page applications. Additionally, they prevent errors with the help of IntelliSense and compile-time type checking. We will investigate tag helpers in *Chapter 12, Working with MVC/Razor Pages*.

Create a new JavaScript file, named `Index.cshtml.js` (you might prefer a different naming style, such as `index.js`; that's fine, as long as you use the same file name in the `abp-script` tag), under the *Pages/Products* folder with the following content:

```
$(function () {
    var l = abp.localization.getResource('ProductManagement');

    var dataTable = $('#ProductsTable').DataTable(
        abp.libs.datatables.normalizeConfiguration({
            serverSide: true,
            paging: true,
            order: [[0, "asc"]],
            searching: false,
            scrollX: true,
            ajax: abp.libs.datatables.createAjax(
                productManagement.products.product.getList),
            columnDefs: [
                /* TODO: Column definitions */
            ]
        })
    );
});
```

ABP allows you to reuse the localization texts in your JavaScript code. In this way, you can define them on the server side and use them on both sides. `abp.localization.getResource` returns a function to localize the values.

ABP simplifies the data table's library configuration and provides built-in integrations:

- `abp.libs.datatables.normalizeConfiguration` is a helper function defined by ABP Framework. It simplifies the data table's configuration by providing conventional default values for missing options.

- `abp.libs.datatables.createAjax` is another helper function that adapts ABP's dynamic JavaScript client proxies to the data table's parameter format.

- `productManagement.products.product.getList` is the dynamic JavaScript proxy function introduced earlier.

Define the columns of the data table inside the `columnDefs` array:

```
{
    title: l('Name'),
    data: "name"
},
{
    title: l('CategoryName'),
    data: "categoryName",
    orderable: false
},
{
    title: l('Price'),
    data: "price"
},
{
    title: l('StockState'),
    data: "stockState",
    render: function (data) {
        return l('Enum:StockState:' + data);
    }
},
{
    title: l('CreationTime'),
```

```
        data: "creationTime",
        dataFormat: 'date'
}
```

Typically, a column definition has a `title` field (display name) and a `data` field. The data field matches the property names in the `ProductDto` class, formatted as **camelCase** (a naming style in which the first letter of each word is capitalized, except for the first word; it is commonly used in the JavaScript language).

The `render` option can be used to finely control how to show the column data. We are providing a function to customize the rendering of the stock state column.

On this page, we've used some localization keys. We should define them in the localization resource. Open the en.json file in the *Localization/ProductManagement* folder of the *ProductManagement.Domain.Shared* project, and add the following entries at the end of the `texts` section:

```
"Name": "Name",
"CategoryName": "Category name",
"Price": "Price",
"StockState": "Stock state",
"Enum:StockState:0": "Pre-order",
"Enum:StockState:1": "In stock",
"Enum:StockState:2": "Not available",
"Enum:StockState:3": "Stopped",
"CreationTime": "Creation time"
```

You can run the web application again to see the product data table in action:

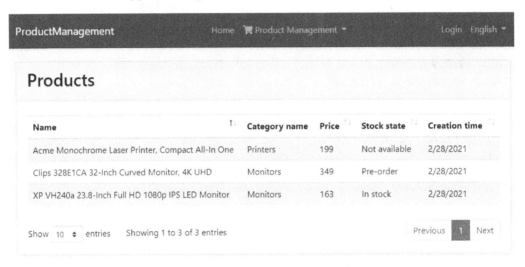

Figure 3.12 – The Products data table

We've created a fully working page that lists the products with paging and sorting support. In the next sections, we will add functionality to create, edit, and delete the products.

Creating products

In this section, we will create the necessary functionality to add a new product. A product should have a category. So, we should select a category while adding a new product. We will define new application service methods to get categories and create products. In the UI section, we will use ABP's dynamic form feature to automatically generate the product creation form, based on a C# class.

Application service contracts

Let's start by adding two new methods to the `IProductAppService` interface:

```
Task CreateAsync(CreateUpdateProductDto input);
Task<ListResultDto<CategoryLookupDto>> GetCategoriesAsync();
```

We will use the `GetCategoriesAsync` method to show a drop-down list of categories on product creation. We've introduced two new DTOs, and we should define them.

`CreateUpdateProductDto` is used to create and update products (we will reuse it in the *Editing products* section). Define it in the *Products* folder of the *ProductManagement.Application.Contracts* project:

```
using System;
using System.ComponentModel.DataAnnotations;
namespace ProductManagement.Products
{
    public class CreateUpdateProductDto
    {
        public Guid CategoryId { get; set; }
        [Required]
        [StringLength(ProductConsts.MaxNameLength)]
        public string Name { get; set; }
        public float Price { get; set; }
        public bool IsFreeCargo { get; set; }
        public DateTime ReleaseDate { get; set; }
        public ProductStockState StockState { get; set; }
    }
}
```

Next, define a `CategoryLookupDto` class in the *Categories* folder of the *ProductManagement.Application.Contracts* project:

```
using System;
namespace ProductManagement.Categories
{
    public class CategoryLookupDto
    {
        public Guid Id { get; set; }
        public string Name { get; set; }
    }
}
```

We've created the contracts, so now we can implement the contracts in the application layer.

Application service implementation

Implement the `CreateAsync` and `GetCategoriesAsync` methods in
`ProductAppService` (in the *ProductManagement.Application* project), as shown
in the following code block:

```
public async Task CreateAsync(CreateUpdateProductDto input)
{
    await _productRepository.InsertAsync(
        ObjectMapper.Map<CreateUpdateProductDto, Product>
(input)
    );
}


public async Task<ListResultDto<CategoryLookupDto>>
        GetCategoriesAsync()
{
    var categories = await _categoryRepository.GetListAsync();
    return new ListResultDto<CategoryLookupDto>(
        ObjectMapper
        .Map<List<Category>, List<CategoryLookupDto>>
(categories)
    );
}
```

Here, _categoryRepository is a type of `IRepository<Category,
Guid>` service. You inject it just as you did for `_productRepository` earlier.
I think the method implementations are pretty simple, and there is no need for
additional explanation.

We've used object mapping in two places, and now we have to define the mapping
configuration. Open the `ProductManagementApplicationAutoMapperProfile.
cs` file in the *ProductManagement.Application* project, and add the following code:

```
CreateMap<CreateUpdateProductDto, Product>();
CreateMap<Category, CategoryLookupDto>();
```

This code sets up the AutoMapper configuration for the object mapping.

> **Automated Tests**
>
> I will not show any more automated tests in this chapter; however, I have added them to the solution. You can check the source code in the GitHub repository.

Now, we can go and consume these methods from the UI layer.

UI

Create a new `CreateProductModal.cshtml` Razor Page under the *Pages/Products* folder of the *ProductManagement.Web* project. Open the `CreateProductModal.cshtml.cs` file, and change the `CreateProductModalModel` class using the following code:

```
using System.Linq;
using System.Threading.Tasks;
using Microsoft.AspNetCore.Mvc;
using Microsoft.AspNetCore.Mvc.Rendering;
using ProductManagement.Products;
namespace ProductManagement.Web.Pages.Products
{
    Public class CreateProductModalModel:
        ProductManagementPageModel
    {
        [BindProperty]
        public CreateEditProductViewModel Product { get;
set; }

        public SelectListItem[] Categories { get; set; }

        private readonly IProductAppService
_productAppService;

        public CreateProductModalModel(
            IProductAppService productAppService)
        {
            _productAppService = productAppService;
        }

        public async Task OnGetAsync()
```

```
        {
            // TODO
        }

        public async Task<IActionResult> OnPostAsync()
        {
            // TODO
        }
    }
}
```

Here, `ProductManagementPageModel` is a base class defined in the startup template. You can inherit it to create `PageModel` classes. `Categories` will be used to show a list of the categories in a drop-down list. `[BindProperty]` is a standard ASP.NET Core attribute to bind the post data to the `Product` property on an HTTP Post request. We are injecting the `IProductAppService` interface to use the methods defined earlier.

We've already used `CreateEditProductViewModel`, so we need to define it. Define it in the same folder as `CreateProductModal.cshtml`:

```
using ProductManagement.Products;
using System;
using System.ComponentModel;
using System.ComponentModel.DataAnnotations;
using Volo.Abp.AspNetCore.Mvc.UI.Bootstrap.TagHelpers.Form;
namespace ProductManagement.Web.Pages.Products
{
    public class CreateEditProductViewModel
    {
        [SelectItems("Categories")]
        [DisplayName("Category")]
        public Guid CategoryId { get; set; }
        [Required]
        [StringLength(ProductConsts.MaxNameLength)]
        public string Name { get; set; }
        public float Price { get; set; }
        public bool IsFreeCargo { get; set; }
        [DataType(DataType.Date)]
```

```
        public DateTime ReleaseDate { get; set; }
        public ProductStockState StockState { get; set; }
    }
}
```

SelectItems tells us that the CategoryId property will be selected from the Categories list. We will reuse this class in the edit modal dialog. That's why I named it CreateEditProductViewModel.

> **DTOs versus ViewModels**
>
> It might seem unnecessary to define the view model (CreateEditProductViewModel) since it is very similar to the DTO (CreateUpdateProductDto). However, it has just a few more attributes. These attributes can be easily added to the DTO, and we can reuse the DTO on the view side. It is up to your design decision, and you can do it. However, I think it is better practice to separate each concern, considering these classes have different purposes and evolve in different directions over time. For example, the [SelectItems("Categories")] attribute refers to the Razor Page model, and it has no meaning in the application layer.

Now, we can implement the OnGetAsync method in the CreateProductModalModel class:

```
public async Task OnGetAsync()
{
    Product = new CreateEditProductViewModel
    {
        ReleaseDate = Clock.Now,
        StockState = ProductStockState.PreOrder
    };

    var categoryLookup =
        await _productAppService.GetCategoriesAsync();
    Categories = categoryLookup.Items
        .Select(x => new SelectListItem(x.Name,
x.Id.ToString()))
            .ToArray();
}
```

We are creating the `Product` class with default values, then filling the `Categories` list using the product application service. `Clock` is a service provided by ABP Framework to get the current time without dealing with time zones and Local/UTC times. We use it instead of `DateTime.Now`. This will be explained in *Chapter 8, Using the Features and Services of ABP*.

We can implement `OnPostAsync`, as shown in the following code block:

```
public async Task<IActionResult> OnPostAsync()
{
    await _productAppService.CreateAsync(
        ObjectMapper

.Map<CreateEditProductViewModel,CreateUpdateProductDto>
(Product)
    );
    return NoContent();
}
```

Since we are mapping `CreateEditProductViewModel` to `CreateProductDto`, we need to define the mapping configuration. Open the `ProductManagementWebAutoMapperProfile` class in the *ProductManagement. Web* project, and change the content using the following code block:

```
public class ProductManagementWebAutoMapperProfile : Profile
{
    public ProductManagementWebAutoMapperProfile()
    {
        CreateMap<CreateEditProductViewModel,
CreateUpdateProductDto>();
    }
}
```

This class defines the object mappings for the AutoMapper library.

We've completed the C# side of the product creation UI.

Now we can start to build the UI markup and JavaScript code. To do this, open the `CreateProductModal.cshtml` file, and change the content as follows:

```
@page
@using Microsoft.AspNetCore.Mvc.Localization
```

```
@using ProductManagement.Localization
@using Volo.Abp.AspNetCore.Mvc.UI.Bootstrap.TagHelpers.Modal
@model ProductManagement.Web.Pages.Products.
CreateProductModalModel
@inject IHtmlLocalizer<ProductManagementResource> L
@{
    Layout = null;
}
<abp-dynamic-form abp-model="Product"
                  asp-page="/Products/CreateProductModal">
    <abp-modal>
        <abp-modal-header title="@L["NewProduct"].Value"></abp-
modal-header>
        <abp-modal-body>
            <abp-form-content />
        </abp-modal-body>
        <abp-modal-footer buttons="@(AbpModalButtons.
Cancel|AbpModalButtons.Save)"></abp-modal-footer>
    </abp-modal>
</abp-dynamic-form>
```

Here, `abp-dynamic-form` automatically creates the form elements based on the C# model class. `abp-form-content` is where the form elements are rendered. `abp-modal` is used to create a modal dialog.

You can use the standard Bootstrap HTML elements and ASP.NET Core's bindings to create form elements. However, ABP's Bootstrap and dynamic form tag helpers simplify the UI code a lot. We will cover ABP tag helpers in *Chapter 12, Working with MVC/Razor Pages*.

We've completed the product creation modal code. Now, we will add a **New Product** button to the products page to open that modal. Open the `Index.cshtml` file in the *Pages/Products* folder, and change the `abp-card-header` section as follows:

```
<abp-card-header>
    <abp-row>
        <abp-column size-md="_6">
            <abp-card-title>@L["Menu:Products"]</abp-card-
title>
        </abp-column>
```

```
<abp-column size-md="_6" class="text-end">
    <abp-button id="NewProductButton"
                text="@L["NewProduct"].Value"
                icon="plus"
                button-type="Primary"/>
</abp-column>
    </abp-row>
</abp-card-header>
```

I've added 2 columns where each column has a `size-md="_6"` attribute (that is half of the 12-column Bootstrap grid). Then, I placed a button on the right-hand side by keeping the card title on the left-hand side.

Following this, I added the following code to the end of the `Index.cshtml.js` file (right before the final `});` code part):

```
var createModal = new abp.ModalManager(abp.appPath +
'Products/CreateProductModal');
createModal.onResult(function () {
    dataTable.ajax.reload();
});
$('#NewProductButton').click(function (e) {
    e.preventDefault();
    createModal.open();
});
```

`abp.ModalManager` is used to manage modal dialogs on the client side. Internally, it uses Twitter Bootstrap's standard modal component but abstracts many details by providing a simple API. `createModal.onResult()` is a callback that is called when the modal is saved. `createModal.open();` is used to open the modal dialog.

Finally, we need to define some localization texts in the `en.json` file in the *Localization/ ProductManagement* folder of the *ProductManagement.Domain.Shared* project:

```
"NewProduct": "New Product",
"Category": "Category",
"IsFreeCargo": "Free Cargo",
"ReleaseDate": "Release Date"
```

You can run the web application again and try to create a new product:

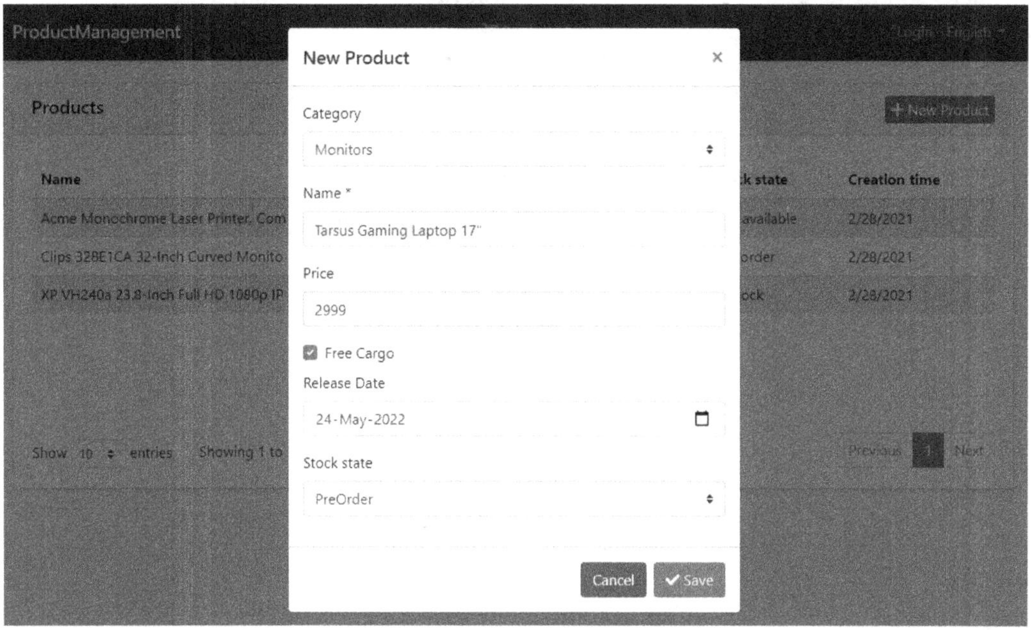

Figure 3.13 – The New Product modal

ABP has automatically created the form fields based on the C# class model. Localization and validation also work automatically by reading the attributes and using the conventions. Try to leave the name field empty and save the modal to see an example of the validation error mesage. We will cover the validation and localization topics in more detail in *Chapter 12, Working with MVC/Razor Pages*.

We can now create products on the UI. Now, let's see how to edit products.

Editing products

Editing a product is similar to adding a new product. This time, we need to get the product to edit and prepare the edit form.

Application service contracts

Let's start by defining two new methods for the IProductAppService interface:

```
Task<ProductDto> GetAsync(Guid id);
Task UpdateAsync(Guid id, CreateUpdateProductDto input);
```

The first method will be used to obtain the product data by ID. We are reusing `CreateUpdateProductDto` (which was defined earlier) in the `UpdateAsync` method.

We haven't introduced a new DTO, so we can go straight to the implementation.

Application service implementation

Implementing these new methods is pretty simple. Add the following methods to the `ProductAppService` class:

```
public async Task<ProductDto> GetAsync(Guid id)
{
    return ObjectMapper.Map<Product, ProductDto>(
        await _productRepository.GetAsync(id)
    );
}

public async Task UpdateAsync(Guid id, CreateUpdateProductDto
input)
{
    var product = await _productRepository.GetAsync(id);
    ObjectMapper.Map(input, product);
}
```

The `GetAsync` method uses `productRepository.GetAsync` to get the product from the database and returns it by mapping it to a `ProductDto` object. The `UpdateAsync` method gets the product and maps the given input properties to the product's properties. In this way, we overwrite the product properties with new values.

For this example, we don't need to call `_productRepository.UpdateAsync` because EF Core has a change tracking system. ABP's **Unit of Work** system automatically saves the changes at the end of the request if it doesn't throw an exception. We will cover the Unit of Work system in *Chapter 6, Working with the Data Access Infrastructure*.

The application layer is now complete. In the next section, we'll create a product editing UI.

UI

Create a new EditProductModal.cshtml Razor Page under the *Pages/Products* folder of the *ProductManagement.Web* project. Open EditProductModal.cshtml.cs, and change the content using the following code:

```
using System;
using System.Linq;
using System.Threading.Tasks;
using Microsoft.AspNetCore.Mvc;
using Microsoft.AspNetCore.Mvc.Rendering;
using ProductManagement.Products;
namespace ProductManagement.Web.Pages.Products
{
    public class EditProductModalModel :
ProductManagementPageModel
    {
        [HiddenInput]
        [BindProperty(SupportsGet = true)]
        public Guid Id { get; set; }

        [BindProperty]
        public CreateEditProductViewModel Product { get; set; }
        public SelectListItem[] Categories { get; set; }

        private readonly IProductAppService _productAppService;

        public EditProductModalModel(IProductAppService
productAppService)
        {
            _productAppService = productAppService;
        }

        public async Task OnGetAsync()
        {
            // TODO
        }
```

```
public async Task<IActionResult> OnPostAsync()
{
    // TODO
}
}
}
```

The `Id` property will be a hidden field in the form. It should also support an HTTP GET request since a GET request opens this modal, and we need the product's ID to prepare the edit form. The `Product` and `Categories` properties are similar to the create modal. We are also injecting the `IProductAppService` interface into the constructor.

We can implement the `OnGetAsync` method, as shown in the following code block:

```
public async Task OnGetAsync()
{
    var productDto = await _productAppService.GetAsync(Id);
    Product = ObjectMapper.Map<ProductDto,
CreateEditProductViewModel>(productDto);

    var categoryLookup = await
_productAppService.GetCategoriesAsync();
    Categories = categoryLookup.Items
        .Select(x => new SelectListItem(x.Name, x.Id.
ToString()))
        .ToArray();
}
```

First, we are getting the product (`ProductDto`) to edit. We are converting it into `CreateEditProductViewModel`, which is then used on the UI to create the edit form. Then, we are getting the categories to select on the form, as we did earlier for the creation form.

We've mapped `ProductDto` to `CreateEditProductViewModel`, so now we need to define the mapping configuration in the `ProductManagementWebAutoMapperProfile` class (in the *ProductManagement. Web* project) just like we've done previously:

```
CreateMap<ProductDto, CreateEditProductViewModel>();
```

The `OnPostAsync` method is simple; we call the `UpdateAsync` method by converting `CreateEditProductViewModel` into `CreateUpdateProductDto`:

```
public async Task<IActionResult> OnPostAsync()
{
    await _productAppService.UpdateAsync(Id,
        ObjectMapper.Map<CreateEditProductViewModel,
CreateUpdateProductDto>(Product)
    );
    return NoContent();
}
```

Now we can switch to `EditProductModal.cshtml`, and change its content as follows:

```
@page
@using Microsoft.AspNetCore.Mvc.Localization
@using ProductManagement.Localization
@using Volo.Abp.AspNetCore.Mvc.UI.Bootstrap.TagHelpers.Modal
@model ProductManagement.Web.Pages.Products.
EditProductModalModel
@inject IHtmlLocalizer<ProductManagementResource> L
@{
    Layout = null;
}
<abp-dynamic-form abp-model="Product"
                  asp-page="/Products/EditProductModal">
    <abp-modal>
        <abp-modal-header title="@Model.Product.Name"></abp-
modal-                header>
        <abp-modal-body>
            <abp-input asp-for="Id" />
            <abp-form-content/>
        </abp-modal-body>
        <abp-modal-footer buttons="@(AbpModalButtons.
Cancel|AbpModalButtons.Save)"></abp-modal-footer>
    </abp-modal>
</abp-dynamic-form>
```

This page is very similar to `CreateProductModal.cshtml`. I just added the `Id` field to the form (as a hidden input) to store the `Id` property of the product being edited.

Finally, we can add an **Edit** action to open that modal from the products data table. Open the `Index.cshtml.js` file, and add a new `ModalManager` object on top of the `dataTable` initialization code:

```
var editModal = new abp.ModalManager(abp.appPath + 'Products/
EditProductModal');
```

Then, add a new column definition as the first item in the `columnDefs` array within the `dataTable` initialization code:

```
{
    title: l('Actions'),
    rowAction: {
        items:
            [
                {
                    text: l('Edit'),
                    action: function (data) {
                        editModal.open({ id: data.record.id });
                    }
                }
            ]
    }
},
```

This code adds a new **Actions** column to the data table and adds an **Edit** action, which opens the edit modal with a click. `rowAction` is a special option provided by ABP Framework. It is used to add one or more actions for a row in the table.

Finally, add the following code after the `dataTable` initialization code:

```
editModal.onResult(function () {
    dataTable.ajax.reload();
});
```

This code refreshes the data table after saving a product edit dialog, so we can see the latest data on the table. The final UI looks similar to *Figure 3.14*:

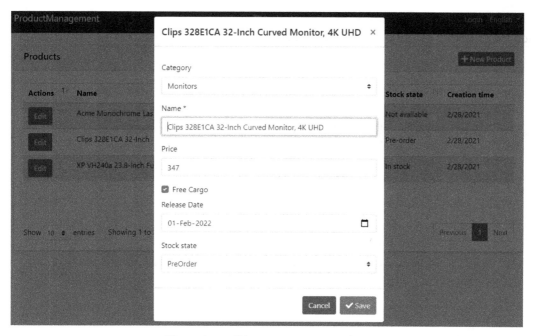

Figure 3.14 – Editing an existing product

We can now see the products, create new products, and edit existing ones. The final section will add a new action to delete an existing product.

Deleting products

Deleting a product is pretty simple compared to the create or edit actions since, in this case, we don't need to build a form. First, add a new method to the `IProductAppService` interface:

```
Task DeleteAsync(Guid id);
```

Then, implement it in the `ProductAppService` class:

```
public async Task DeleteAsync(Guid id)
{
    await _productRepository.DeleteAsync(id);
}
```

We can now add a new action to the product data table. Open `Index.cshtml.js`, and add the following definition just after the **Edit** action (in the `rowAction.items` array):

```
{
    text: l('Delete'),
    confirmMessage: function (data) {
        return l('ProductDeletionConfirmationMessage',
data.record.name);
    },
    action: function (data) {
        productManagement.products.product
            .delete(data.record.id)
            .then(function() {
                abp.notify.info(l('SuccessfullyDeleted'));
                dataTable.ajax.reload();
            });
    }
}
```

Here, `confirmMessage` is a function that is used to get a confirmation from the user before executing the action. The `productManagement.products.product.delete` function is dynamically created by ABP Framework, as explained earlier. In this way, you can directly call server-side methods in your JavaScript code. We are just passing the current record's ID. It returns a promise so that we can register a callback to the `then` function. Finally, we use `abp.notify.info` to send a notification to inform the user, then refresh the data table.

We've used some localization texts, so we need to add the following lines to the localization file (the `en.json` file in the *Localization/ProductManagement* folder of the *ProductManagement.Domain.Shared* project):

```
"ProductDeletionConfirmationMessage": "Are you sure to delete
this book: {0}",
"SuccessfullyDeleted": "Successfully deleted!"
```

You can run the web project again to see the result:

Figure 3.15 – The Delete action

The **Edit** button automatically turns into an **Action** drop-down button since we now have two actions. When you click on the **Delete** action, you get a confirmation message to delete the product:

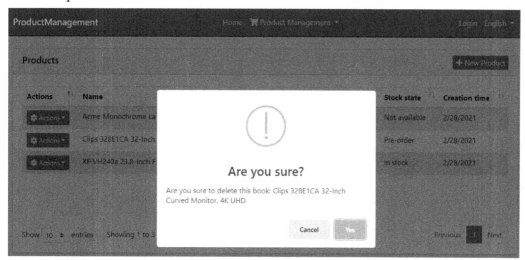

Figure 3.16 – The delete confirmation message

If you click on the **Yes** button, you will see a notification on the page, and the data table will be refreshed.

Implementing product deletion was pretty simple. ABP's built-in features helped us by implementing common patterns such as client-to-server communication, confirmation dialogs, and UI notifications.

Notice that the `Product` entity has been inherited from the `FullAuditedAggregateRoot` class that made it Soft-Delete. Check the database after deleting a product. You will see that it was not really deleted, but the `IsDeleted` field is set to `true`. Setting `IsDeleted` to `true` makes the product entity soft-deleted (that is, deleted logically but not physically). The next time you query products, deleted products are automatically filtered and not included in the query result. That's done by the data filtering system of ABP Framework and will be covered in *Chapter 6, Working with the Data Access Infrastructure*.

Summary

In this chapter, we created a fully working CRUD page. We went through all layers of the application and have seen the fundamental approaches of ABP-based application development.

You were introduced to many different concepts, such as entities, repositories, database mapping and migrations, automated tests, API controllers, dynamic JavaScript proxies, object to object mapping, Soft-Delete, and more. If you're building a serious software solution, you will use all of them, with ABP or not. ABP is a full-stack application framework that helps you implement these concepts with best practices. It provides the necessary infrastructure to make your daily development easier.

You might not understand all of the details at this point. That's not a problem because the purpose of the remaining chapters is to deep dive into these concepts and show their details and different use cases.

This example application was relatively simple. It doesn't contain any important business logic because I've introduced many concepts and tried to keep the application simple to focus on these concepts rather than business complexities. I've ignored authorization in this example. Authorization will be explained in *Chapter 7, Exploring Cross-Cutting Concerns*.

Demonstrating an example application with real-world complexity is not easy in a book. However, I've prepared a complete reference application with real-world qualities and complexities for the readers of this book. The reference application is open source and available on GitHub. Additionally, it is a live application, so you can try it directly.

The next chapter will introduce that reference application and show the reference solution's functionalities, layers, and code structure. The remaining chapters frequently refer to the source code of that application.

4
Understanding the Reference Solution

In the previous chapter, we built a simple full-stack web application that is used to manage products with categories. We've seen a typical flow of developing applications with ABP Framework. You are now ready to create your own application with the basic features. In the next chapters, you will better understand the ABP features and create more advanced applications.

Giving examples with real-world complexities in a book is not very easy. Having reflected on that, we've prepared a complete, real-world reference application built with ABP Framework: *EventHub*. It is open source and freely available on GitHub.

The EventHub solution is thought of as a live system that is available on `openeventhub.com`. You can just try it out to explore it. We've established the **continuous integration/ continuous development (CI/CD)** pipelines, and we are updating the website as we develop it and get contributions from the community. Feel free to see its source code, submit bug reports or feature requests, or even send your pull requests to contribute! As the name suggests, this is an open platform.

This book is the only source of documentation that explains the EventHub solution because we've mainly prepared it for the readers of this book. I will refer to that solution in the next chapters of the book, especially in *Part 3, Implementing Domain-Driven Design*.

In this chapter, we will investigate the EventHub solution in the following sections:

- Introducing the application
- Understanding the architecture
- Running the solution

Technical requirements

You can clone or download the source code of the EventHub project from GitHub, at `https://github.com/volosoft/eventhub`.

If you want to run the solution in your local development environment, you need to have an **integrated development environment (IDE)**/editor (such as Visual Studio) to build and run ASP.NET Core solutions. You also need to have **Docker** installed on your computer. You can download and install **Docker Desktop** for the development environment by following the documentation at `https://docs.docker.com/ get-docker`.

Introducing the application

EventHub is a platform that is used to create organizations to organize events. You create events, either online or in person, then people register them. The following screenshot is taken from the **Home** page of the openeventhub.com website:

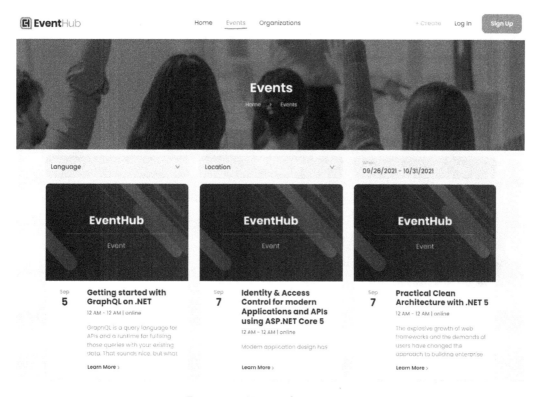

Figure 4.1 – EventHub Home page

You can explore upcoming **Events** section on the **Home** page. Click on an event for details and register for the event. You get an email notification before the event starts or the event time changes.

Here is another screenshot from the **Create New Event** page:

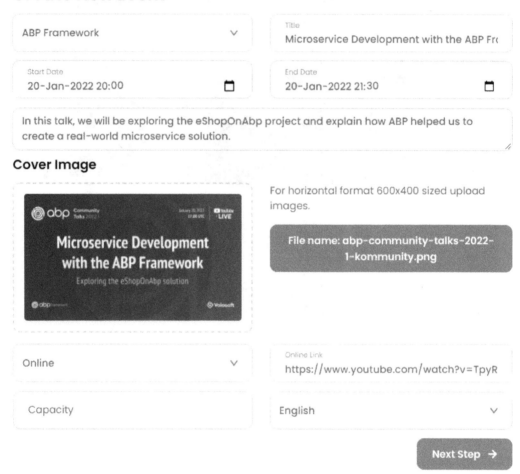

Figure 4.2 – Create New Event page

You can select one of your owned organizations on this page, set a **Title**, time, and description, pick a **Cover Image**, and determine other details about the event you are organizing.

If you want to learn more, please register at openeventhub.com and explore the platform. In this book, I want to talk about the technical details rather than the application's features. Let's begin with the big picture and understand the solution's architecture.

Understanding the architecture

Here is an overall diagram of the applications inside the solution:

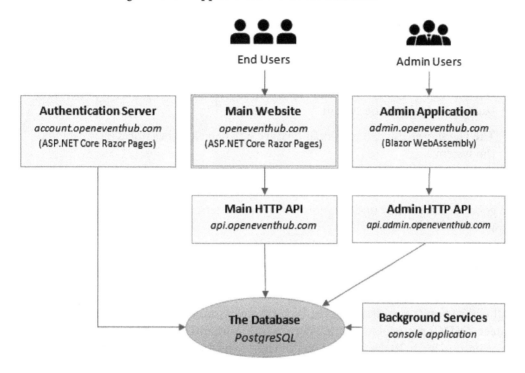

Figure 4.3 – Applications of the EventHub solution

There are six applications and one database shown in *Figure 4.3*, and more information on them is provided here:

- **Authentication Server**: This application is used for logging in, registering, and managing the user account. It is based on ABP's standard **Account** module, which is based on the `IdentityServer` library. It is a **single sign-on** (SSO) server, which means that if you log in to one of the applications, you are then logged in to all of the applications (and vice versa, meaning that if you log out of one of the applications, you are logged out of all of the applications). That is an **ASP.NET Core Razor Pages** application, and it directly connects to **The Database**.

- **Main Website**: This is an essential website (`www.openeventhub.com`) of the platform used by **End Users** to create new events and register for events. It is an **ASP.NET Core Razor Pages** application that uses the **Main HTTP API** as the backend.

- **Admin Application**: This application allows **Admin Users** to manage organizations, events, and the system. It uses the **Admin HTTP API** for all the operations, which is a **Blazor WebAssembly** application that runs in the browser.

- **Main HTTP API**: Exposes **HyperText Transfer Protocol (HTTP) application programming interfaces (APIs)** to be consumed by the main website.

- **Admin HTTP API**: Exposes HTTP APIs to be consumed by the admin application.

- **Background Services**: A **console application** that runs background workers and background jobs of the system.

- **The Database**: This is a relational **PostgreSQL** database that stores all the data in the system.

Since it is a distributed system, it uses **Redis** as the distributed cache server.

It is a good idea to start by understanding the authentication flow to then understand the system.

Authentication flow

As mentioned in the previous section, the **Authentication Server** is an SSO server used to authenticate users and clients. **Main Website** and **Admin Application** use the **OpenID Connect (OIDC)** protocol to redirect users to the **Authentication Server** when users want or need to log in to the application. The following diagram shows the login process:

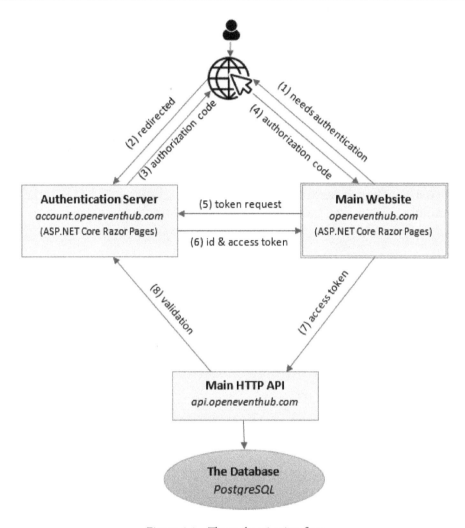

Figure 4.4 – The authentication flow

In *Figure 4.4*, the logic process occurs in the following order:

- Whenever a user wants to visit a page that requires authentication **(1)** or a user explicitly clicks to the login link, the **Main Website** redirects the user **(2)** to the **Authentication Server**.

- **Authentication Server** has a login page so that users can enter a username and password or register as a new user. Once the login process is done, the user is redirected back to the **Main Website** with an authorization code **(3)** and **(4)**.

- The **Main Website** then performs a token request **(5)** to the server using the obtained authorization code.

- **Authentication Server** returns an **identifier** (**ID**) token (contains some user information such as username, ID, email, and so on) and an access token (**6**).

- The **Main Website** stores the access token in a cookie so that it can be obtained in the next requests. In the next requests, it gets the access token from the cookie and adds it to the HTTP request header while performing HTTP requests to the **Main HTTP API** application (**7**).

- The **Main HTTP API** application validates the access token (**8**) and authorizes the request.

The **Main Website** uses cookies to store the access token, as mentioned. On the other hand, the **Admin (Blazor WebAssembly) Application** stores the access token in the local storage of the browser and adds it to the HTTP request header in every request to the server.

All that process is done by ABP's `Account` and `IdentityServer` modules with some configurations in the applications. I won't show the detailed configuration here to keep this chapter focused on the overall solution structure and architecture (check the source code for more details).

In the next section, we will explore the EventHub .NET solution and the projects inside it.

Exploring the solution

The EventHub .NET solution consists of several projects, grouped by the application type and shown in the following screenshot:

Figure 4.5 – EventHub .NET solution in Visual Studio

The solution contains a single domain layer with two application layers and corresponding HTTP API and **user interface (UI)** layers. Two applications use the single domain layer but they have different application logic, so they are separated. We will return to this topic (multiple application layers) in the *Dealing with multiple applications* section of *Chapter 9, Understanding Domain-Driven Design*.

Let's begin explaining the projects with the core part, the common folder. That folder contains common libraries and services, as outlined here:

- The EventHub.Domain project is the domain layer that contains the entities, domain services, and other domain objects. The EventHub.Domain.Shared project contains constants and some other classes, which are shared among all layers and applications in the solution.

- The EventHub.EntityFrameworkCore project contains the code that defines **Entity Framework Core (EF Core)** DbContext, mappings, database migrations, repository implementations, and other code related to EF Core.

- The `EventHub.DbMigrator` project is a console application that you can run to apply pending database migrations and seed the initial data (such as the admin user/ role and their permissions). It is suitable to use in development and production environments.

- The `EventHub.BackgroundServices` project is another console application that runs background workers and jobs on the system and should always be running.

The www folder contains the **Main Website** (`www.openeventhub.com`) application's components, listed as follows:

- The `EventHub.Application` project is the application layer that contains the implementation of the application services, while the `EventHub.Application.Contracts` project includes the application service interfaces and **data transfer objects** (**DTOs**) that are shared with the UI layer.

- The `EventHub.HttpApi` project contains the API controllers that are consumed by the UI (web) layer. The controllers in that project are simple wrappers around the application services.

- The `EventHub.HttpApi.Host` project hosts the HTTP API layer. In this way, the hosting logic is separated from the project that contains the API controllers (which makes it possible to reuse the `EventHub.HttpApi` project as a library).

- The `EventHub.HttpApi.Client` project is a library that can be referenced by a .NET application to consume the API controllers easily. The UI (web) layer uses that project to call the HTTP APIs. This project uses ABP's dynamic C# proxy feature, which will be covered in *Chapter 14, Building HTTP APIs and Real-Time Services*. In this way, we don't need to deal with HTTP clients and low-level details to call HTTP APIs from the UI layer.

- The `EventHub.Web` project is the UI layer of the application. That is a typical Razor Pages application that renders the **HyperText Markup Language** (**HTML**) in the server. It has no database connection but uses the **Main HTTP API** application for all operations.

- The `EventHub.Web.Theme` project is a custom theme for the application. ABP has a theming system that you can use to build your own themes and reuse them in any application. The `EventHub.Web` project uses this theme. Theming system will be covered in *Part 4, User Interface and API Development*.

The `admin` folder contains the admin application that is used by the users who maintain the system, and is explained in more detail here:

- The `EventHub.Admin.Application` project is the application layer of the admin side that contains the implementation of the application services, while the `EventHub.Admin.Application.Contracts` project includes the application service interfaces and DTOs that are shared with the UI layer.

- The `EventHub.Admin.HttpApi` project contains the API controllers that are consumed by the UI (web) layer.

- The `EventHub.Admin.HttpApi.Host` project hosts the HTTP API layer. In this way, the hosting logic is separated from the project that contains the API controllers.

- The `EventHub.Admin.HttpApi.Client` project is a library that can be referenced by a .NET application to consume the API controllers easily. The UI (web) layer uses that project to call the HTTP APIs. This project uses ABP's dynamic C# proxy feature, which will be covered in *Chapter 14, Building HTTP APIs and Real-Time Services*. In this way, we don't need to deal with HTTP clients and low-level details to call HTTP APIs from the UI layer.

- The `EventHub.Admin.Web` project is the UI layer of the application. That is a **Blazor WebAssembly** application that runs in the browser and performs HTTP API calls to the server.

Finally, the `account` folder contains the **Authentication Server**, which contains a single project, `EventHub.IdentityServer`, that is used by other applications to authenticate the users.

I've explained all the projects in the solution, in brief. It is also important to understand the relations and dependencies between the projects.

Project dependencies

Separating the solution into multiple projects makes it possible to have multiple applications on runtime while sharing the code base between applications where it is necessary.

In the next sections, I will show the dependency graph of each application so that you can understand how the code base is organized. We begin with the **Main Website**, the essential application.

Main Website

Remember that the **Main Website** is the application that is used by the **End Users**: www. openeventhub.com. I won't use the *EventHub* prefix for project names anymore since all the projects have the same prefix, and there is no need to repeat it everywhere. The following diagram shows the project dependencies, beginning from the root project of the application—Web:

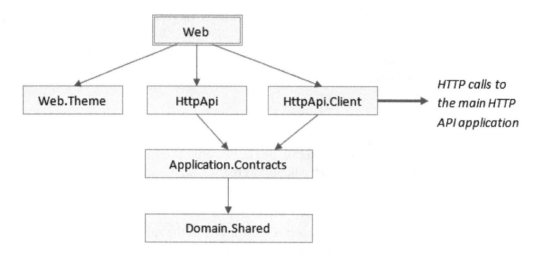

Figure 4.6 – Main website project dependencies

The Web project depends on Web.Theme, which implements the EventHub application's UI theme. Web.Theme is a separate project because it is reused from the **Authentication Server** application. That is an example of reusing a UI theme between multiple applications.

The Web project also depends on the HttpApi project. In this way, the HTTP API controllers become available in the web application, and we can consume these APIs from the client (JavaScript) code. However, when you call an HTTP API controller of this application, the request is redirected to the **Main HTTP API** (backend) by using the HttpApi.Client package. Notice that both the HttpApi and HttpApi. Client project reference the Application.Contacts project. API controllers in the HttpApi project use the application service interfaces, while the HttpApi.Client package implements these interfaces (using ABP's dynamic C# proxy system, which will be explored in *Chapter 14, Building HTTP APIs and Real-Time Services*) to perform remote HTTP calls to the **Main HTTP API** application. So, this application becomes a proxy for the direct API calls between the client (JavaScript) and the HTTP API server. The actual implementations of the application service interfaces run in the **Main HTTP API** application, which will be explained in the next section.

Main HTTP API

Main HTTP API is used by the **Main Website** as the backend API. It runs the application and domain logic of the application and is deployed to `api.openeventhub.com`. The following diagram shows the root `HttpApi.Host` project and its direct and indirect dependencies:

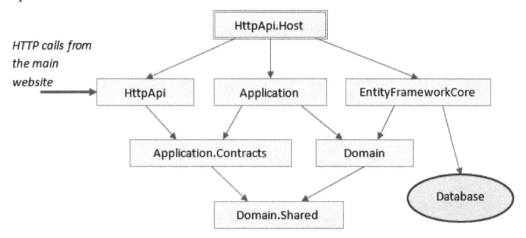

Figure 4.7 – Main HTTP API project dependencies

By referencing (adding a project dependency to) the `HttpApi` project (which includes the API controllers), we can respond to HTTP API calls. The API controllers use the application service interfaces defined in the `Application.Contracts` project. These interfaces are implemented by the `Application` project. That is why we need to reference the `Application` project from the `HttpApi.Host` project. The `Application` project uses the `Domain` project to perform the business logic of the application.

The `HttpApi.Host` project also references the `EntityFrameworkCore` project since we need a data layer on runtime. The `EntityFrameworkCore` project maps entities to the tables in the database, and implements the repositories defined in the `Domain` project.

Notice that the `Application.Contracts` project (and the `Domain.Shared` project, indirectly) is shared by the client application, the **Main Website**, so they can rely on the same application service interfaces to communicate.

We have now explored the **Main Website** application components. The next section continues from the admin side.

Admin application

The admin application is a Blazor WebAssembly application that runs on the browser and is accessed using the following **Uniform Resource Locator** (**URL**): admin. openeventhub.com. It is used by the users who maintain the system. This application has a different set of APIs, UI pages, authorization rules, caching requirements, and so on. Hence, we've created a different application and HTTP API layers for that application. Nevertheless, it shares the same domain layer, so it uses the same domain logic and the same database.

Let's start from the following diagram of the frontend (Blazor WebAssembly) application:

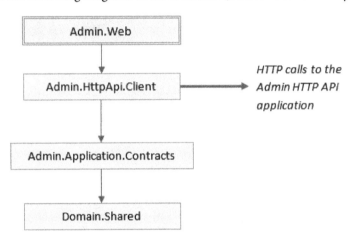

Figure 4.8 – Admin website project dependencies

This diagram is simple compared to previous ones. The Admin.Web project (which is the Blazor WebAssembly application) references the Admin.HttpApi.Client project because it needs to call remote HTTP APIs. ABP's dynamic C# client proxy system (covered in *Chapter 14, Building HTTP APIs and Real-Time Services*) makes it possible to use the application service interfaces in the Blazor WebAssembly application to consume Admin HTTP APIs on the server easily. The Admin.HttpApi.Client project depends on the Admin.Application.Contracts project (which internally depends on the Domain.Shared project) to be able to use the application service interfaces defined in that project.

Admin HTTP API

The **Admin HTTP API** application is used by the admin website as a backend API. It runs the application and domain logic of the admin application and is deployed to https:// admin-api.openeventhub.com. The following diagram shows the root Admin. HttpApi.Host project and its direct and indirect dependencies:

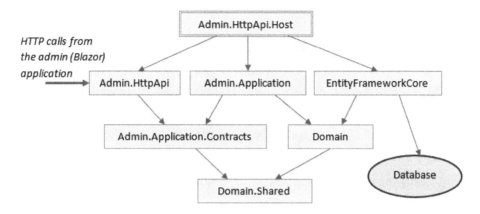

Figure 4.9 – Admin HTTP API project dependencies

The diagram is very similar to the diagram of the **Main HTTP API** application. The difference is that the **Admin Application** has different HTTP API and application layers. However, it uses the same **Domain** and database integration (EntityFrameworkCore) layers to share the same core domain rules and the same **Database**. I will return to that topic in the *Dealing with multiple applications* section of *Chapter 9, Understanding Domain-Driven Design*.

All the applications use the **Authentication Server** application as an SSO server, discussed in the next section.

Authentication Server

The **Authentication Server**'s root project is the IdentityServer project and has the dependencies shown in the following diagram:

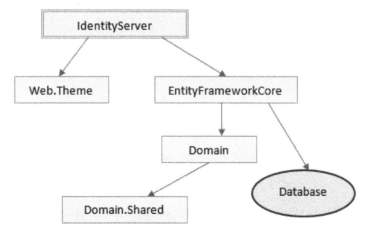

Figure 4.10 – Authentication Server project dependencies

The IdentityServer project has a reference to the Web.Theme project, which is the UI theme that is shared with the **Main Website**. It also references the EntityFrameworkCore project to be able to use the **Database**. By referencing the EntityFrameworkCore project, we also have indirect references to the Domain and Domain.Shared projects.

The next section shows the dependencies of the final application in the solution.

Background Services

The BackgroundServices project has the dependencies shown in the following diagram:

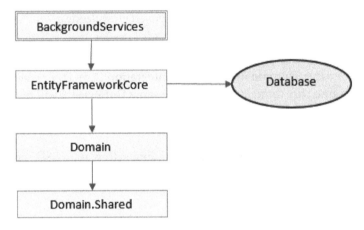

Figure 4.11 – BackgroundServices project dependencies

The BackgroundServices project uses the EntityFrameworkCore project so that it can work with the **Database**. It can also use the **Domain** objects (entities, domain services) to perform background tasks.

We've explored all the projects in the solution. Now, we are ready to run them in our local development environment.

Running the solution

If you want to run the solution in your local environment, follow the steps in the next sections.

Cloning the GitHub repository

First of all, you need to clone the GitHub repository on your local computer. The repository is located at `https://github.com/volosoft/eventhub` and can be cloned using the following command (which requires Git tools to be installed):

```
git clone https://github.com/volosoft/eventhub.git
```

Alternatively, navigate to `https://github.com/volosoft/eventhub`, click the **Code** button, and then click **Download ZIP**, as shown in the following screenshot:

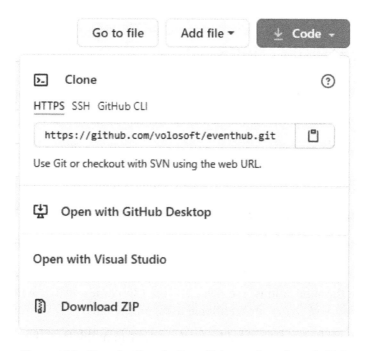

Figure 4.12 – Downloading the EventHub repository from GitHub

You should extract the ZIP file into an empty folder.

Running the infrastructure

The EventHub solution needs **Redis** and **PostgreSQL** servers. The repository contains `docker-compose` files in the `etc/docker` folder. If you have the Docker installed on your computer, you can execute the `up.ps1` file in that folder to run these servers. If you can't use PowerShell on your computer, you can just open it in a text editor, copy the content, and execute it in a command-line terminal in the `etc/docker` directory. In the first run, it may take a few minutes to download the Docker images. If you don't want to use Docker, you need to install **Redis** and **PostgreSQL** servers on your computer.

Opening the solution

The cloned or downloaded repository contains an `EventHub.sln` file in the root folder. If you want to develop or debug the solution, open it in Visual Studio or another .NET-compatible IDE.

Creating a database

The solution has a console application named `EventHub.DbMigrator` that is shown in *Figure 4.5*. Run this application (for Visual Studio, right-click on it and choose **Set as the startup project**, then hit *Ctrl + F5*). It will create a database and seed some initial data.

Running the applications

We are now ready to start the actual applications. You can run the projects in the following order (for Visual Studio, right-click on each project, select **Set as the startup project**, then hit *Ctrl + F5*):

- `EventHub.IdentityServer`
- `EventHub.HttpApi.Host`
- `EventHub.Web`
- `EventHub.Admin.HttpApi.Host`
- `EventHub.Admin.Web`
- `EventHub.BackgroundServices`

To log in to one of the applications, use `admin` as the username and `1q2w3E*` as the password. Of course, you can create additional users on the UI.

Notice that Visual Studio has some problems when you run multiple applications. Sometimes, a previously run application may stop. In this case, run the stopped applications again. However, Microsoft's `Tye` project makes it much easier to run multiple applications.

Using the Tye project

If you don't need to develop or debug the solution but just want to run it, you can use the Microsoft Tye project to run it without opening an IDE. Tye is a .NET global tool used to run such distributed applications easily with a simple configuration file. The EventHub solution was configured to run with Tye. All you need to do is to install Tye and run it.

Before using Tye, you still need to run the infrastructure (see the *Running the infrastructure* section), then create a database using the EventHub.DbMigrator application. If you haven't done this before, open a command-line terminal in the src/EventHub.DbMigrator directory and run the following command:

```
dotnet run
```

After the database is ready, you can install Tye using the following command in a command-line terminal:

```
dotnet tool install -g Microsoft.Tye
```

At the time of writing this book, the Tye project was still in preview. You may need to specify the latest preview version (you can find this on NuGet, at https://www.nuget.org/packages/Microsoft.Tye). For example, see the following code snippet:

```
dotnet tool install -g Microsoft.Tye --version "0.10.0-
alpha.21420.1"
```

Check https://github.com/dotnet/tye/blob/main/docs/getting_started.md to learn how to install Tye.

Tye requires Docker to be installed on your computer. If you haven't installed it yet, you also need to do so. After all the installation is done, you can run the following command to start the applications (it is suggested to close the IDE first if it is open):

```
tye run
```

It will take some time on the first run. Once it is complete, you can open a browser and navigate to `http://127.0.0.1:8000` to open the **Tye Dashboard**, which you can see in the following screenshot:

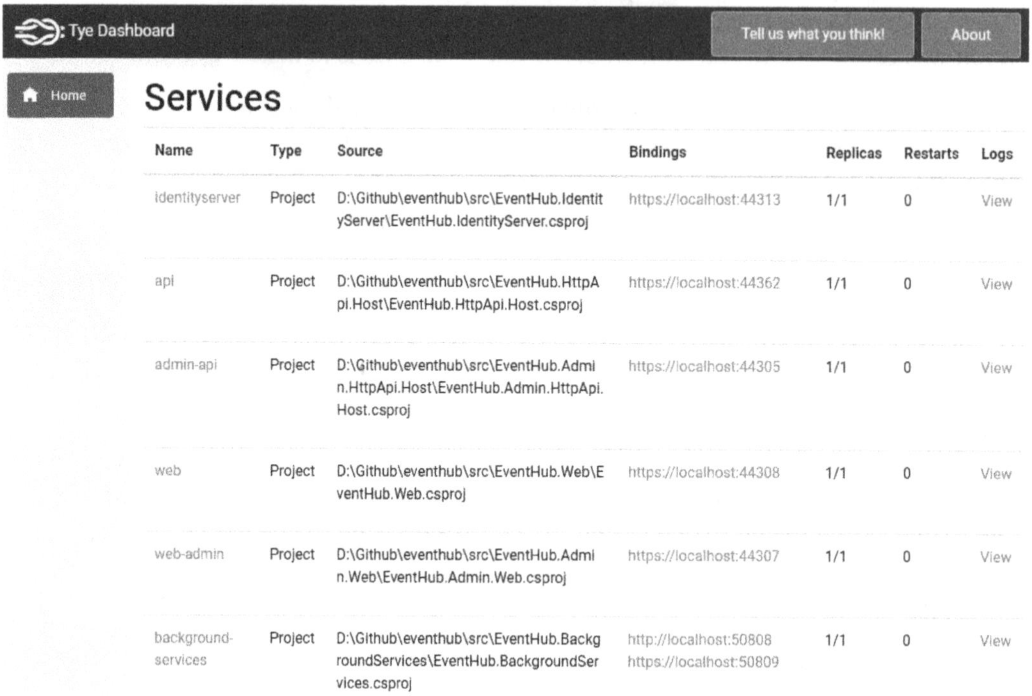
Tye Dashboard
Tell us what you think!
About
Home
Services
Name
Type
Source
Bindings
Replicas
Restarts
Logs
identityserver
Project
D:\Github\eventhub\src\EventHub.IdentityServer\EventHub.IdentityServer.csproj
https://localhost:44313
1/1
0
View
api
Project
D:\Github\eventhub\src\EventHub.HttpApi.Host\EventHub.HttpApi.Host.csproj
https://localhost:44362
1/1
0
View
admin-api
Project
D:\Github\eventhub\src\EventHub.Admin.HttpApi.Host\EventHub.Admin.HttpApi.Host.csproj
https://localhost:44305
1/1
0
View
web
Project
D:\Github\eventhub\src\EventHub.Web\EventHub.Web.csproj
https://localhost:44308
1/1
0
View
web-admin
Project
D:\Github\eventhub\src\EventHub.Admin.Web\EventHub.Admin.Web.csproj
https://localhost:44307
1/1
0
View
background-services
Project
D:\Github\eventhub\src\EventHub.BackgroundServices\EventHub.BackgroundServices.csproj
http://localhost:50808
https://localhost:50809
1/1
0
View

Figure 4.13 – The Tye Dashboard

The **Tye Dashboard** is used to view the applications and their **Logs** in real time. You can click links on the **Bindings** column for any application to open. `web` is the **Main Website** of the system.

Tye is a handy tool when you have a solution with multiple applications that need to run together. You can also configure `dotnet watch` for a project so that it is automatically reloaded (or hotloaded with .NET 6.0) when you change it. See Microsoft's documentation to learn more about this: `https://github.com/dotnet/tye/tree/main/docs`.

Summary

EventHub is a complete, real-world quality application built on ABP Framework. It is not just an example but also a live project published on openeventhub.com and actively developed on https://github.com/volosoft/eventhub. Feel free to send bug reports, feature requests, and pull requests.

In this chapter, my purpose was not to explain the code base in detail. I explained the overall architecture and structure of the solution so that you can understand how to explore the code base and run the solution. The next chapters will refer to that solution while introducing some ABP features and concepts.

EventHub is a good example of a system that was built with multiple applications. It is also a clear example to understand the purpose of ABP's layering model and how to reuse these layers in different applications.

You may not understand all the details of the EventHub solution now because we haven't explained the module system, database integrations, dynamic C# client proxies, and all the other ABP features yet. The chapters in the next part will explore the fundamental building blocks of ABP Framework and the ASP.NET Core framework so that you will start to understand all the details.

In the next chapter, we will explore the basic building blocks of ASP.NET Core and ABP Framework to understand how an application is configured and initialized.

Part 2: Fundamentals of ABP Framework

In this part, you will learn about the infrastructure provided by ABP Framework to achieve common software development requirements. You will see how ABP saves you time by helping you to implement the **DRY (Don't Repeat Yourself)** principle by automating common tasks with conventions.

In this part, we include the following chapters:

5
Exploring the ASP.NET Core and ABP Infrastructure

Both ASP.NET Core and ABP Framework provide many building blocks and features for modern application development. This chapter will explore the most basic building blocks so that you can understand how an application is configured and initialized.

We will start with the ASP.NET Core `Startup` class to understand why we need a modular system and how ABP provides a modular way to configure and initialize an application. Then we will explore the ASP.NET Core dependency injection system and ABP's way of automating dependency injection registration with predefined rules. We will continue by looking at configuration and the options pattern to learn ASP.NET Core's way of configuring the options of ASP.NET Core and other libraries.

Here are all the topics we'll cover in this chapter:

- Understanding modularity
- Using the dependency injection system
- Configuring an application
- Implementing the options pattern
- Logging

Technical requirements

If you want to follow and try the examples, you need to have installed an IDE/editor (such as Visual Studio) to build the ASP.NET Core projects.

You can download the code examples from the following GitHub repository: `https://github.com/PacktPublishing/Mastering-ABP-Framework`.

Understanding modularity

Modularity is a design technique for breaking down the functionalities of a large software into smaller parts and allowing each part to communicate with the others through standardized interfaces as needed. Modularity has the following main benefits:

- It reduces complexity when every module is designed to be isolated from the other modules, and inter-module communications are well defined and limited.
- It provides flexibility when you design modules to be loosely coupled. You can refactor or even replace a module in the future.
- It allows re-using modules across applications when you design them to be application-independent.

Most enterprise software systems are designed to be modular. However, implementing modularity is not easy, and the plain ASP.NET Core doesn't help much. One of ABP Framework's main goals is to provide infrastructure and tooling to develop truly modular systems. We will cover modular application development in *Chapter 15, Working with Modularity*, but this section introduces the basics of ABP modules.

The Startup class

Before defining a module class, it is best to remember the Startup class in ASP.NET Core to understand the need for module classes. The following code block shows a Startup class in a simple ASP.NET Core application:

```
public class Startup
{
    public void ConfigureServices(IServiceCollection
services)
    {
        services.AddMvc();
        services.AddTransient<MyService>();
    }

    public void Configure(
        IApplicationBuilder app, IWebHostEnvironment env)
    {
        app.UseRouting();
        if (env.IsDevelopment())
        {
            app.UseDeveloperExceptionPage();
        }
        app.UseEndpoints(endpoints =>
        {
            endpoints.MapControllers();
        });
    }
}
```

The ConfigureServices method is used to configure other services and register new services to the dependency injection system. The Configure method, on the other hand, is used to configure the **ASP.NET Core request pipeline** that processes the HTTP requests through middleware components.

Once you have the `Startup` class, you typically register it in the `Program.cs` file while configuring the host builder so that it works on application startup:

```
public class Program
{
    public static void Main(string[] args)
    {
        CreateHostBuilder(args).Build().Run();
    }
    public static IHostBuilder CreateHostBuilder(string[]
args) =>
        Host.CreateDefaultBuilder(args)
            .ConfigureWebHostDefaults(webBuilder =>
            {
                webBuilder.UseStartup<Startup>();
            });
}
```

These code parts are already included in ASP.NET Core's startup templates, so you normally don't write them manually.

The problem with the `Startup` class is that it is unique. That means you have only a single point to configure and initialize all your application services. However, in a modular application, you expect that every module configures and initializes the services related to that particular module. Also, it is typical that a module uses or depends on other modules, so the modules should be configured and initialized in the correct order. That is where ABP's module definition class comes into play.

Defining a module class

An ABP module is a group of types (such as classes or interfaces) developed and shipped together. It is an assembly (a *project* in Visual Studio) with a module class derived from `AbpModule`. The module class is responsible for configuring and initializing that module and configures any dependent modules if necessary.

Here is a simple module definition class for an SMS sending module:

```
using Microsoft.Extensions.DependencyInjection;
using Volo.Abp.Modularity;
namespace SmsSending
{
```

```
public class SmsSendingModule : AbpModule
{
    public override void ConfigureServices(
        ServiceConfigurationContext context)
    {
        context.Services.AddTransient<SmsService>();
    }
}
```

Every module can override the ConfigureServices method in order to register its services to the dependency injection system and configure the other modules. The module in this example registers SmsService to a dependency injection system with a transient lifetime. I've written this example to show the same registration code done in the Startup class in the previous section. However, most of the time, you don't need to register your services manually, thanks to ABP Framework's conventional registration system explained in the *Using the dependency injection system* section of this chapter.

The AbpModule class defines the OnApplicationInitialization method that is executed after the service registration phase is complete and the application is ready to run. With this method, you can execute any operation you need to perform on application startup. For example, you can initialize a service:

```
public class SmsSendingModule : AbpModule
{
    //...
    public override void OnApplicationInitialization(
        ApplicationInitializationContext context)
    {
        var service = context.ServiceProvider
            .GetRequiredService<SmsService>();
        service.Initialize();
    }
}
```

In this code block, we are using context.ServiceProvider to request a service from the dependency injection system and initialize the service. We can request services because the dependency injection system is ready at this point.

You can also think of the OnApplicationInitialization method as the Configure method of the Startup class. So, you can build the ASP.NET Core request pipeline here. However, you typically configure the request pipeline in the startup module, as explained in the next section.

Module dependencies and the startup module

A business application generally consists of more than one module, and ABP Framework allows you to declare dependencies between modules. An application should always have a **startup module**. The startup module can have dependencies on some modules, and these modules can have dependencies on some other modules, and so on.

The following diagram shows a simple module dependency graph:

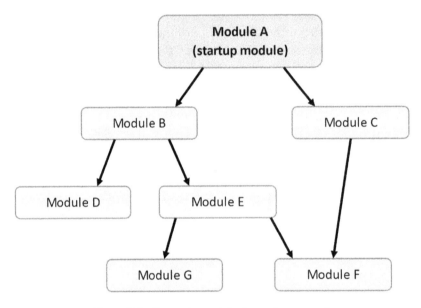

Figure 5.1 – Example module dependency graph

ABP respects module dependencies and initializes modules based on the dependency graph. If module A depends on module B, then module B is always initialized before module A. That allows module A to use, set, change, or override configurations and services defined by module B.

For the example graph in *Figure 5.1*, the module initialization will be in the following order: G, F, E, D, B, C, A. You don't have to know the exact initialization order; only know that if your module depends on module *X*, then module *X* is initialized before your module.

Defining a module dependency is declared with the [DependsOn] attribute of a module:

```
[DependsOn(typeof(ModuleB), typeof(ModuleC))]
public class ModuleA : AbpModule
{
}
```

In the preceding code block, ModuleA depends on ModuleB and ModuleC by declaring the [DependsOn] attribute.

For an ASP.NET Core application, the startup module (ModuleA in this example) is responsible for setting up the ASP.NET Core request pipeline:

```
[DependsOn(typeof(ModuleB), typeof(ModuleC))]
public class ModuleA : AbpModule
{
    //...
    public override void OnApplicationInitialization(
        ApplicationInitializationContext context)
    {
        var app = context.GetApplicationBuilder();
        var env = context.GetEnvironment();

        app.UseRouting();
        if (env.IsDevelopment())
        {
            app.UseDeveloperExceptionPage();
        }
        app.UseEndpoints(endpoints =>
        {
            endpoints.MapControllers();
        });
    }
}
```

With this code block, we've built the same ASP.NET Core request pipeline built previously in the *The Startup class* section. `context.GetApplicationBuilder()` and `context.GetEnvironment()` are just shortcuts to obtain the standard `IApplicationBuilder` and `IWebHostEnvironment` services from the dependency injection system.

Then, we can use this module in the `Startup` class of ASP.NET Core to integrate ABP Framework with ASP.NET Core:

```
public class Startup
{
    public void ConfigureServices(IServiceCollection
services)
    {
        services.AddApplication<ModuleA>();
    }

    public void Configure(IApplicationBuilder app)
    {
        app.InitializeApplication();
    }
}
```

The `services.AddApplication()` method is defined by ABP Framework to configure the modules. It basically executes the `ConfigureServices` methods of all the modules by respecting the order of module dependencies. The `app. InitializeApplication()` method is also defined by ABP Framework; similarly, it executes the `OnApplicationInitialization` methods of all the modules by respecting the order of module dependencies.

The `ConfigureServices` and `OnApplicationInitialization` methods are the most commonly used methods in a module class; there are more methods explained in the next section.

Module lifecycle methods

The `AbpModule` class defines useful methods that you can override to execute code on application startup and shutdown. We saw `ConfigureServices` and `OnApplicationInitialization` in the previous section; here is a list of all the lifecycle methods:

- `PreConfigureServices`: This method is called before the `ConfigureServices` method. It allows you to write code to be executed before `ConfigureServices` of the depended-upon modules.

- `ConfigureServices`: This is the main method to configure the module and register services, as explained in the previous section.

- `PostConfigureServices`: This method is called after the `ConfigureServices` method of all the modules (including the modules depending on your module), so you may perform a final configuration.

- `OnPreApplicationInitialization`: This method is called before the `OnApplicationInitialization` method. In this stage, you can resolve services from dependency injection.

- `OnApplicationInitialization`: This method allows your module to configure the ASP.NET Core request pipeline and initialize your services, as explained in the previous section.

- `OnPostApplicationInitialization`: This method is called in the initialization phase.

- `OnApplicationShutdown`: You can implement your module's shutdown logic if necessary.

`Pre...` and `Post...` methods (such as `PreConfigureServices` and `PostConfigureServices`) have the same purpose as the original method. They are rarely used and provide a way to perform some configuration/initialization code that works before or after all other modules.

> **Asynchronous Lifecycle Methods**
>
> The lifecycle methods explained in this section are synchronous. At the time of writing this book, the ABP Framework team was working to introduce asynchronous lifecycle methods with ABP Framework version 5.1. You can see `https://github.com/abpframework/abp/pull/10928` for details.

As explained before, a module class mainly contains code to register and configure services related to that module. In the next section, we will see how to register services with ABP Framework.

Using the dependency injection system

Dependency injection is a technique for obtaining a class's dependencies. It separates creating a class from using that class.

Assume that we have a `UserRegistrationService` class that uses `SmsService` to send a verification SMS, as shown in the following code block:

```
public class UserRegistrationService
{
    private readonly SmsService _smsService;

    public UserRegistrationService(SmsService smsService)
    {
        _smsService = smsService;
    }

    public async Task RegisterAsync(
        string username,
        string password,
        string phoneNumber)
    {
        //...save user in the database
        await _smsService.SendAsync(
            phoneNumber,
            "Your verification code: 1234"
        );
    }
}
```

Here, `SmsService` has been obtained using the **Constructor-Injection pattern**. Using the Constructor-Injection pattern practically means that we define parameters in the constructor of our class, then we let the dependency injection system instantiate the class's dependencies and pass them to the constructor of our class. We then assign these object instances to the fields in our class to use later in our methods. `SmsService`, in this example, is used in the `RegisterAsync` method to send a verification code after saving the user into the database.

ASP.NET Core natively provides a dependency injection infrastructure, and ABP leverages this infrastructure rather than using a third-party dependency injection framework. Once you register all the services to the dependency injection system, any service can constructor-inject the dependent services without dealing with creating them (and their dependencies).

The most important thing you should consider while designing your services is the service lifetime.

Service lifetime

ASP.NET Core offers different lifetime options on service registration, so we should select a lifetime for every service. There are three lifetimes in ASP.NET Core:

- **Transient**: Transient services are created whenever you inject them. Every time you request/inject the service, a new instance is created.

- **Scoped**: Scoped services are created per scope. This is generally considered by request lifetime, as each HTTP request creates a new scope in ASP.NET Core. You share the same instance in the same scope and get a different instance in different scope.

- **Singleton**: A singleton service has only a single instance in an application. All the requests and clients use the same instance. The object is created the first time you request it. Then the same object instance is reused in the subsequent requests.

The following module registers two services, one as transient and the other as a singleton:

```
public class MyModule : AbpModule
{
    public override void ConfigureServices(
        ServiceConfigurationContext context)
    {
        context.Services.AddTransient<ISmsService,
SmsService>();
```

```
        context.Services.AddSingleton<OtherService>();
    }
}
```

`context.Services` is a type of `IServiceCollection`, and all the ASP.NET Core extension methods can be used to register and configure your services manually.

In the first example, `AddTransient<ISmsService, SmsService>()`, I've registered the `SmsService` class with the `ISmsService` interface. In this way, whenever I inject `ISmsService`, the dependency injection system creates an `SmsService` object for me. For the second example, `AddSingleton<OtherService>()`, I've registered `OtherService` as a singleton with the class reference. To use this service, I should inject the `OtherService` class reference.

> **Scoped Dependencies and ASP.NET Core's Dependency Injection Documentation**
>
> As mentioned, scoped services are created per HTTP request for an ASP.NET Core application by default. For non-ASP.NET Core applications, you may need to manage scopes yourself. Please refer to ASP.NET Core's documentation for all the details of the dependency injection system: `https://docs.microsoft.com/en-us/aspnet/core/fundamentals/dependency-injection`.

When you use ABP Framework, you don't have to think so much about service registration, thanks to ABP Framework's conventional and declarative service registration system.

Conventional service registrations

In ASP.NET Core, you should explicitly register all your services to `IServiceCollection`, as shown in the previous section. However, most of these registrations are just repetitive code and can be automated.

ABP automatically registers services for dependency injection for the following types:

- MVC controllers
- Razor page models
- View components
- Razor components

- SignalR hubs
- Application services
- Domain services
- Repositories

All these services are registered with the transient lifetime. So, you don't need to care about service registration for these kinds of classes. If you have another class type, you can use one of the dependency interfaces, or the Dependency attribute as explained in the next sections.

Dependency interfaces

You can implement an ITransientDependency, IScopedDependency, or ISingletonDependency interface to register your service for dependency injection. For example, in this code block, we've registered the service as a singleton, so only one shared instance is created in the application's lifetime:

```
public class UserPermissionCache : ISingletonDependency
{ }
```

Dependency interfaces are easy and the suggested way for most cases, but they are limited compared to the Dependency attribute.

The Dependency attribute

The Dependency attribute provides options for fine control of dependency registration with the following properties:

- Lifetime (enum): The lifetime of the service: Singleton, Transient, or Scoped
- TryRegister (bool): Registers the service only if it's not already registered
- ReplaceServices (bool): Replaces the previous registration if the service is already registered

Here is an example of service registration using the Dependency attribute:

```
using Microsoft.Extensions.DependencyInjection;
using Volo.Abp.DependencyInjection;
namespace UserManagement
{
```

```
    [Dependency(ServiceLifetime.Transient, TryRegister =
true)]
    public class UserPermissionCache
    { }
}
```

Here, I used the [Dependency] attribute with a Transient lifetime and also with the TryRegister option to register the class to the dependency injection system.

> **Dependency Attribute versus Dependency Interfaces**
>
> The Dependency attribute can be used alongside the dependency interfaces introduced in the previous section. The Dependency attribute has a higher priority than the dependency interfaces if it defines the Lifetime property.

Registering a class to the dependency injection system makes it usable in an application. However, a class may be injected with different types of class or interface references, depending on what service types are exposed by that class.

Exposing services

When a class doesn't implement an interface, it can only be injected by the class reference. The UserPermissionCache class in the previous section is used by directly injecting the class type. However, it is common to implement interfaces for services.

Assume that we have an interface to abstract the SMS sending:

```
public interface ISmsService
{
    Task SendAsync(string phoneNumber, string message);
}
```

That is a pretty simple interface that only has a single method to send an SMS. Assume that you want to implement the ISmsService interface by using Azure:

```
public class AzureSmsService : ISmsService,
ITransientDependency
{
    public async Task SendAsync(string phoneNumber, string
message)
    {
        //TODO: ...
```

```
        }
    }
```

The `AzureSmsService` class implements the `ISmsService` and `ITransientDependency` interfaces. The `ITransientDependency` interface is only for registering this service for dependency injection, as explained in the previous section.

You typically want to use the `AzureSmsService` class by injecting the `ISmsService` interface. ABP is smart enough to understand your purpose and automatically registers the `AzureSmsService` class for the `ISmsService` interface. You can consume the `AzureSmsService` class either by injecting the `ISmsService` interface or the `AzureSmsService` class reference. Injecting the `AzureSmsService` class by the `ISmsService` interface is possible by its naming conventions: the `ISmsService` interface is the *default interface* for the `AzureSmsService` class because it ends with the `SmsService` suffix.

Assume that we have a class that implements multiple interfaces as shown in the following code block:

```
public class PdfExporter: IExporter, IPdfExporter, ICanExport,
ITransientDependency
{ }
```

The `PdfExporter` service can be used by injecting the `IPdfExporter` and `IExporter` interfaces or directly with the `PdfExporter` class reference. However, you can't inject it using the `ICanExport` interface because the name of `PdfExporter` doesn't end with `CanExport`.

If you need to change the default behavior, you can use the `ExposeServices` attribute, as shown in the following code block:

```
[ExposeServices(typeof(IPdfExporter))]
public class PdfExporter: IExporter, IPdfExporter, ICanExport,
ITransientDependency
{ }
```

Now, you can use the `PdfExporter` class only by injecting the `IPdfExporter` interface.

> **Question: Should I Define Interfaces for Each Service?**
>
> One potential question you may ask is whether you should define interfaces for your services and inject them using interfaces. ABP doesn't force you to do anything here, and general interface best practices are applicable: define interfaces if you want to loosely couple your services, have multiple implementations of a service, mock in unit tests easily, physically separate interfaces from the implementations (say, we define application service interfaces in the `Application.Contracts` project and implement them in the `Application` project, or we define repository interfaces in the domain layer but implement them in the infrastructure layer), and so on.

We've seen how to register and consume services. Some services or libraries have options, and you may need to configure them before using them. The next two sections explain the standard infrastructure and patterns to configure options provided by such services and libraries.

Configuring an application

ASP.NET Core's **configuration** system provides a convenient way to read key-value-based configurations for applications. It is an extensible system and can read key-value pairs from various resources, such as JSON settings files, environment variables, command-line arguments, and Azure Key Vault.

> **ABP Framework versus ASP.NET Core's Configuration System**
>
> ABP Framework doesn't add a specific feature to ASP.NET Core's configuration system. However, it is essential to understand it to work with ASP.NET Core and ABP Framework properly. I will cover the basics in this book. Please see ASP.NET Core's documentation for a complete reference: `https://docs.microsoft.com/en-us/aspnet/core/fundamentals/configuration`.

Setting the configuration values

The easiest way to set a configuration value is to use the `appsettings.json` file by default. Assume that we are building a service to send SMS using Azure, and we need the following configuration values:

- `Sender`: The sender number shown to the target user
- `ConnectionString`: The connection string of your Azure resouce

We can define these settings in the configuration section of the `appsettings.json` file:

```
{
    ...
    "AzureSmsService": {
        "Sender": "+901112223344",
        "ConnectionString": "..."
    }
}
```

The configuration section name (`AzureSmsService` here) and key names are completely arbitrary. You can set any name as long as you use the same keys in your code.

Once you've set values in the settings file, you can easily read them from your application code.

Reading the configuration values

You can inject and use the `IConfiguration` service whenever you need to read configured values. For example, we can get the Azure configuration to send SMS in the `AzureSmsService` class:

```
using System.Threading.Tasks;
using Microsoft.Extensions.Configuration;
using Volo.Abp.DependencyInjection;
namespace SmsSending
{
    public class AzureSmsService : ISmsService,
ITransientDependency
    {
        private readonly IConfiguration _configuration;

        public AzureSmsService(IConfiguration configuration)
        {
            _configuration = configuration;
        }

        public async Task SendAsync(
            string phoneNumber, string message)
```

```
        {
            string sender = _
configuration["AzureSmsService:Sender"];
            string ConnectionString = _
configuration["AzureSmsService:ConnectionString"];
            //TODO: Use Azure to send the SMS message
        }
    }
}
```

This class gets the configuration values from the `IConfiguration` service, and the `:` notation is used to access values in nested sections. In this example, `AzureSmsService:Sender` is used to get the `Sender` value inside the `AzureSmsService` section.

The `IConfiguration` service is also usable in the `ConfigureServices` of your module:

```
public override void ConfigureServices(
    ServiceConfigurationContext context)
{
    IConfiguration configuration =
context.Services.GetConfiguration();
    string sender =
configuration["AzureSmsService:Sender"];
}
```

This way, you can access the configured values even before the dependency injection registration phase is completed.

The configuration system is a perfect way to configure and get key-value-style settings for your application. However, if you are building a reusable library, the options pattern can be a better way to define type-safe options for your library.

Implementing the options pattern

With the **options pattern**, we use a plain class (sometimes called a **POCO – Plain Old C# Object**) to define a group of related options. Let's begin with how to define, configure, and use the configuration using the options pattern.

Defining an options class

An options class is a simple plain C# class. We can define an options class for the Azure SMS service as shown in the following code block:

```
public class AzureSmsServiceOptions
{
    public string Sender { get; set; }
    public string ConnectionString { get; set; }
}
```

It is a convention to add the Options suffix to options classes. Once you define such a class, any module using this service can configure the options easily.

Configuring the options

As mentioned in the *ABP modules* section, you can configure the services of the dependent modules in the ConfigureServices method of your module. We use the IServiceCollection.Configure extension method to set values for any options class. You can configure AzureSmsServiceOptions as shown in the following code block:

```
[DependsOn(typeof(SmsSendingModule))]
public class MyStartupModule : AbpModule
{
    public override void ConfigureServices(
        ServiceConfigurationContext context)
    {
        context.Services
            .Configure<AzureSmsServiceOptions>(options =>
            {
                options.Sender = "+901112223344";
                options.ConnectionString = "...";
            });
    }
}
```

The `context.Services.Configure` method is a generic method that gets the options class as the generic parameter. It also takes a delegate (an action) to set the option values. In this example, we've configured `AzureSmsServiceOptions` by setting the `Sender` and `ConnectionString` properties in the specified lambda expression.

The `AbpModule` base class provides a `Configure` method as a shortcut of the `context.Services.Configure` method, so you could re-write the code as follows:

```
public override void ConfigureServices(
    ServiceConfigurationContext context)
{
    Configure<AzureSmsServiceOptions>(options =>
    {
        options.Sender = "+901112223344";
        options.ConnectionString = "...";
    });
}
```

We've just replaced the `context.Services.Configure<...>` call with the `Configure<...>` shortcut method.

Configuring the options is simple. Now, we can see how to use the configured values.

> **Multiple Configure Actions**
>
> You can configure the same options multiple times in an application. The same instance is sent to all delegates so that you can change the previously configured values. If multiple modules configure the same value, the last one wins. Remember that modules are initialized by the dependency order.

Using the configured option values

ASP.NET Core provides an `IOptions<T>` interface to inject the options class to read the configured values. We can re-write the `AzureSmsService` class to use `AzureSmsServiceOptions` instead of the `IConfiguration` service, as in the following code block:

```
public class AzureSmsService : ISmsService,
ITransientDependency
{
    private readonly AzureSmsServiceOptions _options;
```

```
    public AzureSmsService(IOptions<AzureSmsServiceOptions>
options)
    {
        _options = options.Value;
    }

    public async Task SendAsync(string phoneNumber, string
message)
    {
        string sender = _options.Sender;
        string ConnectionString = _options.ConnectionString;
        //TODO...
    }
}
```

Notice that we are injecting IOptions<AzureSmsServiceOptions> and using its Value property to obtain the AzureSmsServiceOptions instance. The IOptions<T> interface is defined by the Microsoft.Extensions. Options package and is the standard way to inject an options class. It internally executes all the Configure methods and provides a configured instance of the options class for you. If you directly inject the AzureSmsServiceOptions class by mistake, you get a dependency injection exception. So, always inject as IOptions<AzureSmsServiceOptions>.

We've simply defined, configured, and used the options. What if we want to use the configuration system to set the options defined with the options pattern?

Setting the options via the configuration

The options pattern allows us to set the option values in any way. That means we can use the IConfiguration service to read the application configuration and set the option values. The following code block sets AzureSmsServiceOptions by getting the values from the configuration service:

```
[DependsOn(typeof(SmsSendingModule))]
public class MyStartupModule : AbpModule
{
    public override void ConfigureServices(
        ServiceConfigurationContext context)
```

```
    {
        var configuration =
context.Services.GetConfiguration();
        Configure<AzureSmsServiceOptions>(options =>
        {
            options.Sender =
configuration["AzureSmsService:Sender"];
            options.ConnectionString =
configuration["AzureSmsService:ConnectionString"];
        });
    }
}
```

We are getting the `IConfiguration` interface using `context.Services.GetConfiguration()`, then using the configuration values to set the option values.

However, since this usage is pretty common, there is a shortcut for it. We could re-write the code as shown in the following block:

```
public override void ConfigureServices(
    ServiceConfigurationContext context)
{
    var configuration = context.Services.GetConfiguration();
    Configure<AzureSmsServiceOptions>(
        configuration.GetSection("AzureSmsService"));
}
```

With this usage, the `Configure` method gets a configuration section instead of a delegate action. It automatically matches the configuration keys with the properties of the options class by naming conventions. This code does not affect the options if the `AzureSmsService` section is not defined in the configuration.

The options pattern gives more flexibility to the application developers; they may set these options from `IConfiguration` or any other source they like.

> **Tip: Set Options from the Configuration by Default**
>
> If you are building a reusable module, it is a good practice to set your options from the configuration wherever possible. That is, you can write the preceding code into your module. In this way, an application developer can directly configure their module from the `appsettings.json` file.

ASP.NET Core and ABP options

ASP.NET Core and ABP Framework intensively use the options pattern for their configuration options.

The following example shows configuring an option in ABP Framework:

```
Configure<AbpAuditingOptions>(options =>
{
    options.IgnoredTypes.Add(typeof(ProductDto));
});
```

AbpAuditingOptions is defined by the audit logging system of ABP Framework. We are adding a type, ProductDto, to be ignored on audit logging.

The next example shows configuring an option in ASP.NET Core:

```
Configure<MvcOptions>(options =>
{
    options.RespectBrowserAcceptHeader = true;
});
```

MvcOptions is defined by ASP.NET Core to customize the behavior of the ASP.NET Core MVC framework.

Complex Types in the Options Classes

Notice that AbpAuditingOptions.IgnoredTypes is a list of Type, which is not a simple primitive type that you can define in an appsettings.json file. That is one of the benefits of the options pattern: you can define properties with complex types or even action callbacks.

The configuration system and the options pattern provide a convenient way to configure and customize the behavior of the services being used. You can configure ASP.NET Core and ABP Framework and define configuration options for your own services.

The next section explains logging, another fundamental system that you will frequently use in your application code.

Logging

Logging is a common aspect used in every application. ASP.NET Core provides a simple yet efficient logging system. It can be integrated with popular logging libraries such as NLog, Log4Net, and Serilog.

Serilog is a widely used library that provides many options for the log target, including console, text files, and Elasticsearch. ABP startup templates come with the Serilog library pre-installed and configured. It writes logs into a log file in the `Logs` folder of the application. So, you can directly use the logging system in your services. If you need, you can configure Serilog to write logs to different targets. Please refer to Serilog's documentation to configure the Serilog options. Serilog is not a core dependency of ABP Framework. All the configuration is included in the startup template. So, if you like, you can easily change it with another provider.

The `ILogger<T>` interface is used to write logs in ASP.NET Core, where `T` is typically your service type.

Here is an example service that writes logs:

```
public class AzureSmsService : ISmsService,
ITransientDependency
{
    private readonly ILogger<AzureSmsService> _logger;

    public AzureSmsService(ILogger<AzureSmsService> logger)
    {
        _logger = logger;
    }

    public async Task SendAsync(string phoneNumber, string
message)
    {
        _logger.LogInformation(
            $"Sending SMS to {phoneNumber}: {message}");
        //TODO...
    }
}
```

The AzureSmsService class injects the ILogger<AzureSmsService> service in its constructor and uses the LogInformation method to write information-level log text to the logging system.

There are more methods on the ILogger interface to write logs with different severity levels, such as LogError and LogDebug. Please refer to ASP.NET Core's documentation for all details of the logging system: https://docs.microsoft.com/en-us/aspnet/core/fundamentals/logging.

Summary

This chapter has covered the core building blocks of ASP.NET Core and ABP Framework.

You've learned about using the Startup class, configuration system, and options pattern to configure ASP.NET Core and ABP Framework services on application startup and implement your own configuration options when you need them.

ABP offers a modularity system that takes ASP.NET Core's initialization and configuration system one step further to create multiple modules where each module initializes its services and configures its dependencies. In this way, you can split your application into modules to better organize your code base or create modules that can be reused in different applications.

The dependency injection system is the most fundamental infrastructure of an ASP.NET Core application. A service consumes others using the dependency injection system. I've introduced the essential aspects of the dependency injection system and explained how ABP simplifies registering your services.

The next chapter focuses on data access infrastructure, which is an essential aspect of a business application. We will see how ABP Framework standardizes defining entities and using repositories to abstract and perform database operations while automating database connections and transaction management.

6
Working with the Data Access Infrastructure

Almost all business applications use a kind of database system. We typically implement data access logic to read data from and write data to a database. We also need to deal with database transactions to ensure consistency in the data source.

In this chapter, we will learn how to work with the data access infrastructure of ABP Framework, which provides abstractions for data access by implementing **Repository** and **Unit of Work (UoW)** patterns. Repositories provide a standard way to perform common database operations for your **entities**. The UoW system automates database connections and transaction management to ensure a use case (typically, a **HyperText Transfer Protocol (HTTP)** request) is atomic; this means all operations done in the request are successful together or rolled back together in any error.

You will see how to define your entities based on ABP Framework's pre-built base entity classes. Then, you will learn how to insert, update, delete, and query entities in the database using the repositories. You will also understand the UoW system to control transaction scopes in your application.

ABP Framework can work with any database system, while it provides built-in integration packages with **Entity Framework Core (EF Core)** and **MongoDB**. You will learn how to use EF Core with ABP Framework by defining your `DbContext` class, mapping your entities to database tables, implementing your repositories, and deploying different ways of loading related entities when you have an entity. You will also see how to use **MongoDB** as a second database provider option.

This chapter covers ABP's fundamental data access infrastructure with the following topics:

- Defining entities

- Working with repositories

- EF Core integration

- MongoDB integration

- Understanding the UoW system

Technical requirements

If you want to follow and try the examples, you need to have an **integrated development environment** (IDE)/editor installed (for example, Visual Studio) to build ASP.NET Core projects.

You can download the code examples from the following GitHub repository: `https://github.com/PacktPublishing/Mastering-ABP-Framework`.

Defining entities

Entities are the main classes to define your domain model. If you are using a relational database, an entity is generally mapped to a database table. An **object-relational mapper (ORM)**, such as EF Core, provides abstractions to make you feel as though you are working with objects in your application code rather than database tables.

ABP Framework standardizes defining entities by providing some interfaces and base classes. In the next sections, you will learn about ABP Framework's `AggregateRoot` and `Entity` base classes (and their variants), using single **primary keys** (**PKs**) and **composite PKs** (**CPKs**) with these classes and working with **globally unique identifier** (**GUID**) PKs.

AggregateRoot classes

An **aggregate** is a cluster of objects (entities and value objects) bound together by an aggregate root object.

Relational databases do not have a physical aggregate concept. Every entity is related to a separate database table, and an aggregate is spread into more than one table. You define relations with **foreign keys (FKs)**. However, in document/object databases such as MongoDB, an aggregate is saved into a single collection by serializing it as a single document (a **JavaScript Object Notation (JSON)**-like object). The aggregate root is mapped to the collection, and sub-entities are serialized within the aggregate root object. That means sub-entities don't have their collections and are always accessed over the aggregate root.

> **The Aggregate Concept**
>
> We will cover the aggregate concept in *Chapter 10, DDD – The Domain Layer*, with all the details of this. For now, you can think of aggregate roots as the main (root) entities in your domain.

In ABP Framework, you can define main entities and aggregate roots, by deriving from one of the `AggregateRoot` classes. `BasicAggregateRoot` is the most simple class for defining your aggregate roots.

The following example entity class is derived from the `BasicAggregateRoot` class:

```
using System;
using System.Collections.Generic;
using Volo.Abp.Domain.Entities;
namespace FormsApp
{
    public class Form : BasicAggregateRoot<Guid>
    {
        public string Name { get; set; }
        public string Description { get; set; }
        public bool IsDraft { get; set; }
        public ICollection<Question> Questions { get; set; }
    }
}
```

`BasicAggregateRoot` just defines an `Id` property as the PK and takes the PK type as the generic parameter. In this example, the PK type of `Form` is `Guid`. You can use any type as the PK (for example, `int`, `string`, and so on), as long as the underlying database provider supports it.

There are some other base classes to derive your aggregate roots from, as detailed here:

- The `AggregateRoot` class has additional properties to support optimistic concurrency and object extension features.

- `CreationAuditedAggregateRoot` inherits from the `AggregateRoot` class and adds `CreationTime` (`DateTime`) and `CreatorId` (`Guid`) properties to store creation audit information.

- `AuditedAggregateRoot` inherits from the `CreationAuditedAggregateRoot` class and adds `LastModificationTime` (`DateTime`) and `LastModifierId` (`Guid`) properties to store modification audit information.

- `FullAuditedAggregateRoot` inherits from the `AuditedAggregateRoot` class and adds `DeletionTime` (`DateTime`) and `DeleterId` (`Guid`) properties to store deletion audit information. It also adds `IsDeleted` (`bool`) by implementing the `ISoftDelete` interface, which makes the entity soft-delete.

> **Optimistic Concurrency and Object Extension Features**
>
> These topics are not covered in this book. Please check the ABP Framework documentation if you need to use them.

ABP automatically sets auditing properties. We will return to the audit logging and soft-delete topics in *Chapter 8, Using the Features and Services of ABP.*

Entity classes

`Entity` base classes are similar to `AggregateRoot` classes, but they are used for sub-collection entities rather than main (root) entities. For example, the `Form` aggregate root example in the previous section has a collection of questions. The `Question` class is derived from the `Entity` class and is shown in the following code snippet:

```
public class Question : Entity<Guid>
{
    public Guid FormId { get; set; }
    public string Title { get; set; }
```

```
    public bool AllowMultiSelect { get; set; }
    public ICollection<Option> Options { get; set; }
}
```

As with the `AggregateRoot` class, the `Entity` class also defines an `Id` property of a given type. In this example, the `Question` entity also has a collection of options, where `Option` is another entity type.

There are some other pre-defined base entity classes, such as `CreationAuditedEntity`, `AuditedEntity`, and `FullAuditedEntity`. They are similar to the audited aggregate root classes explained in the previous section.

Entities with CPKs

Relational databases support CPKs, whereby your PK consists of a combination of multiple values. Composite keys are especially useful for relation tables with **many-to-many** relations.

Assume that you want to set multiple managers for a form object and add a collection property to the `Form` class, as follows:

```
public class Form : BasicAggregateRoot<Guid>
{
    ...

    public ICollection<FormManager> Managers { get; set; }
}
```

You can then define a `FormManager` class deriving from the non-generic `Entity` class, like this:

```
public class FormManager : Entity
{
    public Guid FormId { get; set; }
    public Guid UserId { get; set; }
    public Guid IsOwner { get; set; }

    public override object[] GetKeys()
    {
        return new object[] {FormId, UserId};
    }
}
```

When you inherit from the non-generic `Entity` class, you have to implement the `GetKeys` method to return an array of the keys. In this way, ABP can use the CPK's value where it is needed. For this example, `FormId` and `UserId` are FKs to other tables, and they build the CPK of the `FormManager` entity.

> **CPKs for Aggregate Roots**
>
> `AggregateRoot` classes also have non-generic versions for CPKs, while it is not so usual to set CPKs for aggregate root entities.

The GUID PK

ABP mostly uses GUIDs as the PK type for pre-built entities. GUIDs are generally compared to auto-increment IDs (such as `int` or `long`, supported by relational databases). Here are some commonly known benefits of using GUIDs as the PK compared to auto-increment keys:

- GUIDs are naturally unique. This works well if you are building distributed systems, using a non-relational database, and need to split or merge tables or integrate external systems.

- GUIDs can be generated on the client side without needing a database round trip. In this way, the client code can know the PK value before saving the entity.

- GUIDs are impossible to guess, so they can be more secure in some cases (for example, if end users see the ID of an entity, they can't find the ID of another entity).

GUIDs also have some disadvantages compared to auto-increment integer values, as follows:

- A GUID is 16 bytes in the storage, higher than `int` (4 bytes) and `long` (8 bytes).
- GUIDs are not sequential by nature, which causes performance problems on clustered indexes. However, ABP offers a solution to that problem.

ABP provides the `IGuidGenerator` service, which generates sequential `Guid` values by default. While it generates sequential values, the values generated by the algorithm are still safe to be universal and random. Generating a sequential value solves the clustered index performance problem.

If you manually set the `Id` value of an entity, always use the `IGuidGenerator` service; never use `Guid.NewGuid()`. If you don't set the `Id` value for a new entity and insert it into the database using a repository, the repository automatically sets it using the `IGuidGenerator` service.

> **GUID versus Auto-Increment**
>
> GUID versus auto-increment PKs is a hot discussion in software development, and there is no clear winner. ABP works with any PK type, so you can make your own choice based on your requirements.

We have now learned the basics of entity definitions and will explore best practices for entities in *Chapter 10*, *DDD – The Domain Layer*. But now, let's continue with the repositories to understand how to work with a database to persist our entities.

Working with repositories

The **Repository pattern** is a common approach to abstract the data access code from the other services of your application. In the next sections, you will learn how to use ABP Framework's generic repositories for your entities to query or manipulate data in the database using pre-defined repository methods. You will also see how to create custom repositories when you need to extend the generic repositories and add your own repository methods to encapsulate your data access logic.

> **Integrating Database Providers**
>
> Database provider integration should be done to use repositories. We will do this in the *EF Core integration* and *MongoDB integration* sections of this chapter.

Generic repositories

Once you have an entity, you can directly inject and use the generic repository for that entity. Here is an example class that uses a repository:

```
using System;
using System.Collections.Generic;
using System.Threading.Tasks;
using Volo.Abp.DependencyInjection;
using Volo.Abp.Domain.Repositories;
namespace FormsApp
```

```
{
    public class FormService : ITransientDependency
    {
        private readonly IRepository<Form, Guid>
_formRepository;

        public FormService(IRepository<Form, Guid>
formRepository)
        {
            _formRepository = formRepository;
        }

        public async Task<List<Form>> GetDraftForms()
        {
            return await _formRepository
                .GetListAsync(f => f.IsDraft);
        }
    }
}
```

In this example, we've injected IRepository<Form, Guid>, the default generic
repository for the Form entity. Then, we've used the GetListAsync method to get
a filtered list of forms from the database. The generic IRepository interface has
two generic parameters: entity type (Form, in this example) and PK type (Guid, in
this example).

> **Repositories for Non-Aggregate Root Entities**
>
> Generic repositories are only available for *aggregate root* entities by default
> because it is a best practice to access aggregates via aggregate root objects.
> However, it is possible to enable generic repositories for other entity types if
> you're using a relational database. We will see the configuration point in the *EF
> Core integration* section.

Generic repositories provide many built-in methods to query, insert, update, and
delete entities.

Inserting, updating, and deleting entities

The following methods can be used to manipulate data in the database:

- `InsertAsync` is used to insert a new entity.

- `InsertManyAsync` is used to insert multiple entities in a single call.

- `UpdateAsync` is used to update an existing entity.

- `UpdateManyAsync` is used to update multiple entities in a single call.

- `DeleteAsync` is used to delete an existing entity.

- `DeleteManyAsync` is used to insert multiple entities in a single call.

> **About Asynchronous Programming**
>
> All repository methods are asynchronous. As a general principle in .NET, it is strongly suggested to write your application code with the `async/await` pattern wherever possible, because in .NET, mixing asynchronous code with synchronous code leads to potential deadlock, timeout, and scalability problems in your application that are not easy to detect and resolve.

If you are using EF Core, these methods may not immediately perform an actual database operation because EF Core uses a change-tracking system. It saves changes only when you call the `DbContext.SaveChanges` method. ABP Framework's UoW system automatically calls the `SaveChanges` method when the current HTTP request successfully finishes. If you want to save changes into the database immediately, you can pass the `autoSave` parameter as `true` to the repository methods.

The following example creates a new `Form` entity and immediately saves it to the database in the `InsertAsync` method:

```
var form = new Form(); // TODO: set the form properties
await _formRepository.InsertAsync(form, autoSave: true);
```

Even if you save changes into the database, the changes may not be visible yet, depending on the transaction isolation level, and will be rolled back if the current transaction fails. We will cover the UoW system in the *Understanding the UoW system* section of this chapter.

The `DeleteAsync` method has an extra overload to delete all the entities satisfying the given condition. The following example deletes all the draft forms in the database:

```
await _formRepository.DeleteAsync(form => form.IsDraft);
```

You can also have a complex condition using logical operators such as && and ||.

> **About Cancellation Tokens**
>
> All repository methods get an optional `CancellationToken` parameter. Cancellation tokens are used to cancel a database operation when needed. For example, if the user closes the browser window, there is no need to continue a long-running database query operation. Most of the time, you don't need to manually pass a cancellation token, since ABP Framework automatically captures and uses the cancellation token from the HTTP request when you don't explicitly pass it.

Querying a single entity

The following methods can be used to fetch a single entity:

- `GetAsync`: Returns a single entity by its `Id` value or a predicate expression. Throws `EntityNotFoundException` if the requested entity was not found.

- `FindAsync`: Returns a single entity by its `Id` value or a predicate expression. Returns `null` if the requested entity was not found.

You should use the `FindAsync` method only if you have custom logic or fallback code, in case the given entity does not exist in the database. Otherwise, use `GetAsync`, which throws a well-known exception that causes the `404` status code to return to the client in an HTTP request.

The following example uses the `GetAsync` method to query a `Form` entity with its `Id` value:

```
public async Task<Form> GetFormAsync(Guid formId)
{
    return await _formRepository.GetAsync(formId);
}
```

Both methods have overloads to pass a predicate expression to query an entity with a given condition. The following example uses the `GetAsync` method to get a `Form` entity with its unique name:

```
public async Task<Form> GetFormAsync(string name)
{
    return await _formRepository
        .GetAsync(form => form.Name == name);
}
```

Use these overloads only if you are expecting a single entity. If your query returns multiple entities, then they throw `InvalidOperationException`. For example, if form names are always *unique* in your system, you can find a form by name, as in this example. However, if your query may return multiple entities, use querying methods that return a list of entities.

Querying a list of entities

Generic repositories provide a lot of options to query entities from the database. The following methods can be used to get a list of entities directly:

- `GetListAsync`: Returns all the entities or a list of entities satisfying the given condition
- `GetPagedListAsync`: Used to query entities by paging

The following code block shows how to get a list of forms filtered by the given name:

```
public async Task<List<Form>> GetFormsAsync(string name)
{
    return await _formRepository
        .GetListAsync(form => form.Name.Contains(name));
}
```

I've passed a lambda expression to the `GetListAsync` method to get all the `Form` entities with the given `name` parameter's value contained in their names.

These methods are simple but limited. If you want to write advanced queries, you can use **Language-Integrated Query (LINQ)** over the repositories.

Using LINQ over the repositories

Repositories provide the `GetQueryableAsync()` method, which returns an `IQueryable<TEntity>` object. You can then use this object to perform LINQ on the entities in the database.

The following example uses a LINQ operation on the `Form` entities to get a list of forms filtered and ordered by their names:

```
public class FormService2 : ITransientDependency
{
    private readonly IRepository<Form, Guid>
_formRepository;
    private readonly IAsyncQueryableExecuter
_asyncExecuter;

    public FormService2(
        IRepository<Form, Guid> formRepository,
        IAsyncQueryableExecuter asyncExecuter)
    {
        _formRepository = formRepository;
        _asyncExecuter = asyncExecuter;
    }

    public async Task<List<Form>>
GetOrderedFormsAsync(string name)
    {
        var queryable = await
_formRepository.GetQueryableAsync();
        var query = from form in queryable
            where form.Name.Contains(name)
            orderby form.Name
            select form;
        return await _asyncExecuter.ToListAsync(query);
    }
}
```

We've first obtained an `IQueryable<Form>` object, then written a LINQ query, and finally executed the query using the `IAsyncQueryableExecuter` service.

An alternative way to write the previous query could be using LINQ extension methods, as follows:

```
var query = queryable
    .Where(form => form.Name.Contains(name))
    .OrderBy(form => form.Name);
```

Having an `IQueryable` object provides you with all the power of LINQ. You can even make joins between multiple `IQueryable` objects obtained from different repositories.

Using the `IAsyncQueryableExecuter` service may seem strange to you. You may expect to call the `ToListAsync` method directly on the query object, like so:

```
return await query.ToListAsync();
```

Unfortunately, `ToListAsync` is an extension method defined by EF Core (or MongoDB, if you are using it) and located inside the `Microsoft.EntityFrameworkCore` NuGet package. If referencing that package from your application layer is not a problem for you, then you can directly use these asynchronous extension methods in your code. However, if you want to keep your application layer ORM-independent, ABP's `IAsyncQueryableExecuter` service provides the necessary abstraction.

IRepository async extension methods

ABP Framework provides all the standard async LINQ extension methods for the `IRepository` interface: `AllAsync`, `AnyAsync`, `AverageAsync`, `ContainsAsync`, `CountAsync`, `FirstAsync`, `FirstOrDefaultAsync`, `LastAsync`, `LastOrDefaultAsync`, `LongCountAsync`, `MaxAsync`, `MinAsync`, `SingleAsync`, `SingleOrDefaultAsync`, `SumAsync`, `ToArrayAsync`, and `ToListAsync`. You can directly use any of these methods on a repository object.

The following example uses the `CountAsync` method to get a count of forms where the name starts with `"A"`:

```
public async Task<int> GetCountAsync()
{
    return await _formRepository
        .CountAsync(x => x.Name.StartsWith("A"));
}
```

Notice that these extension methods are only available on the `IRepository` interface. If you want to use queryable extensions, you should still follow the approach explained in the previous section.

Generic repositories for entities with CPKs

If your entity has a CPK, you can't use the `IRepository<TEntity, TKey>` interface, since it gets a single PK (`Id`) type. In this case, you can use the `IRepository<TEntity>` interface.

For example, you can use `IRepository<FormManager>` to get managers of a given form, as follows:

```
public class FormManagementService : ITransientDependency
{
    private readonly IRepository<FormManager>
_formManagerRepository;

    public FormManagementService(
        IRepository<FormManager> formManagerRepository)
    {
        _formManagerRepository = formManagerRepository;
    }

    public async Task<List<FormManager>>
GetManagersAsync(Guid formId)
    {
        return await _formManagerRepository
            .GetListAsync(fm => fm.FormId == formId);
    }
}
```

In this example, I've used the `IRepository<FormManager>` interface to perform a query for the `FormManager` entities.

> **Repositories for Non-Aggregate Root Entities**
>
> As stated in the *Generic repositories* section of this chapter, you can't use `IRepository<FormManager>` by default, since `FormManager` is not an aggregate root entity. You normally want to get the `Form` aggregate root and access its `Managers` collection to get the form managers. However, if you are using EF Core, you can create default generic repositories for entities that are not aggregate roots. See the *EF Core integration* section to learn how to do this.

One limitation of generic repositories without the TKey generic argument is that they don't have methods that get Id parameters because they can't know the Id type. However, you can still use LINQ to write any type of query you need.

Other generic repository types

You typically want to use the repository interfaces explained in the previous sections since they are the most feature-full repository types. However, there are some more limited repository types that can be useful in some scenarios, such as the following:

- IBasicRepository<TEntity, TPrimaryKey> and IBasicRepository<TEntity> provide fundamental repository methods, but they don't support LINQ and IQueryable functionalities. You can use these repositories if your underlying database provider doesn't support LINQ or you don't want to leak LINQ queries into your application layer. In this case, you probably need to write custom repositories by inheriting from these interfaces and implement your queries with custom methods.

- IReadOnlyRepository<TEntity, TKey>, IReadOnlyRepository<TEntity>, IReadOnlyBasicRepository<Tentity, TKey>, and IReadOnlyBasicRepository<TEntity, TKey> provide methods to fetch data but do not include any methods to manipulate the database.

Generic repository methods are enough for most cases. However, you may still need to add custom methods to your repositories.

Custom repositories

You can create custom repository interfaces and classes to access the underlying database provider **application programming interface (API)**, encapsulate your LINQ expressions, call stored procedures, and so on.

To create a custom repository, first, define a new repository interface. Repository interfaces are defined in the Domain project that comes with the startup template. You can inherit from one of the generic repository interfaces to include the standard methods in your repository interface. The code is illustrated in the following snippet:

```
public interface IFormRepository : IRepository<Form, Guid>
{
    Task<List<Form>> GetListAsync(
        string name,
        bool includeDrafts = false
```

```
    );
}
```

`IFormRepository` inherits from `IRepository<Form, Guid>` and adds a new method to get a list of forms with some filters. You can then inject `IFormRepository` into your services instead of the generic repository and use your custom methods. If you don't want to include the standard repository methods, just derive your interface from the `IRepository` (without any generic argument) interface. This is an empty interface that is used to identify your interface as a repository.

Surely, we must implement the `IFormRepository` interface somewhere in our application. ABP startup templates provide integration projects for the underlying database provider, so we can implement custom repository interfaces in the database integration project. We will implement that interface for EF Core and MongoDB in the next sections.

EF Core integration

Microsoft's EF Core is the de facto ORM for .NET, with which you can work with major database providers, such as SQL Server, Oracle, MySQL, PostgreSQL, and Cosmos DB. It is the default database provider when you create a new ABP solution using the ABP **command-line interface** (**CLI**).

The startup template uses *SQL Server* by default. If you prefer another **database management system** (**DBMS**), you can specify the `-dbms` parameter while creating a new solution, like so:

```
abp new DemoApp -dbms PostgreSQL
```

`SqlServer`, `MySQL`, `SQLite`, `Oracle`, and `PostgreSQL` are directly supported.

> **Other Databases**
>
> You can refer to ABP's documentation to learn about up-to-date supported database options and how to switch to another database provider that the ABP CLI does not support out of the box: `https://docs.abp.io/en/abp/latest/Entity-Framework-Core-Other-DBMS`.

In the next sections, you will learn how to configure the DBMS (although it is already done in the startup template), define a `DbContext` class, and register to the **dependency injection** (**DI**) system. Then, you will see how to map your entities to database tables, using Code First Migrations and creating custom repositories for your entities. Finally, we will explore different ways of loading related data for an entity.

Configuring the DBMS

We use `AbpDbContextOptions` to configure the DBMS in the `ConfigureServices` method of our module. The following example configures using SQL Server as the DBMS:

```
Configure<AbpDbContextOptions>(options =>
{
    options.UseSqlServer();
});
```

Surely, the `UseSqlServer()` method call will be different if you've preferred a different DBMS. We don't need to set the connection string since it is automatically obtained from the `ConnectionStrings:Default` configuration. You can check the `appsettings.json` file in your project to see and change the connection string.

We've configured the DBMS but haven't defined a `DbContext` object, which is necessary to work with the database in EF Core.

Defining DbContext

`DbContext` is the main object in EF Core that you interact with the database. You normally create a class inheriting from `DbContext` to create your own `DbContext`. With ABP Framework, we are inheriting from `AbpDbContext` instead.

Here is an example of a `DbContext` class definition with ABP Framework:

```
using Microsoft.EntityFrameworkCore;
using Volo.Abp.EntityFrameworkCore;
namespace FormsApp
{
    public class FormsAppDbContext :
AbpDbContext<FormsAppDbContext>
    {
        public DbSet<Form> Forms { get; set; }
```

```
public FormsAppDbContext(
    DbContextOptions<FormsAppDbContext> options)
    : base(options)
    {

    }
    }
}
```

`FormsAppDbContext` inherits from `AbpDbContext<FormsAppDbContext>`.
`AbpDbContext` is a generic class and takes the `DbContext` type as a generic parameter.
It also forces us to create a constructor, as shown here. We can then add `DbSet` properties
for our entities. It is essential to add `DbSet` properties since ABP can create default
generic repositories only for the entities with `DbSet` properties defined.

Once we've defined `DbContext`, we should register it with the DI system to use it in our
application.

Registering DbContext with DI

The `AddAbpDbContext` extension method is used to register `DbContext` classes with
the DI system. You can use this method inside the `ConfigureServices` method of
your module (it is inside the `EntityFrameworkCore` project in the startup solution),
as shown in the following code block:

```
public override void ConfigureServices(
    ServiceConfigurationContext context)
{
    context.Services.AddAbpDbContext<FormsAppDbContext>
(options =>
    {
        options.AddDefaultRepositories();
    });
}
```

`AddDefaultRepositories()` is used to enable default generic repositories for your entities related to that `DbContext`. It enables generic repositories only for aggregate root entities by default because, in **domain-driven design** (**DDD**), sub-entities should always be accessed over the aggregate root. You can set the optional `includeAllEntities` parameter to `true` if you want to use repositories for other entity types too, as illustrated here:

```
options.AddDefaultRepositories(includeAllEntities: true);
```

With this option, you can inject the `IRepository` service for any entity in your application code.

> **The includeAllEntities Option in the Startup Template**
>
> The ABP startup template sets the `includeAllEntities` option to `true` because developers working on relational databases are used to querying from all database tables. If you want to apply DDD principles strictly, you should always use the aggregate roots to access sub-entities. In this case, you can remove this option from the `AddDefaultRepositories` method call.

We've seen how to register the `DbContext` class. We can inject and use `IRepository` interfaces for all your entities in your `DbContext` class. However, we should first configure the EF Core mappings for the entities.

Configuring entity mappings

EF Core is an object-to-relational mapper that maps your entities to database tables. We can configure the details of those mappings in two ways, as outlined here:

- Using data annotation attributes on your entity class
- Using Fluent API inside by overriding the `OnModelCreating` method

Using data annotation attributes makes your domain layer EF Core-dependent. If that's not a problem for you, you can simply use these attributes by following EF Core's documentation. In this book, I will use the Fluent API approach.

To use the Fluent API approach, you can override the `OnModelCreating` method in your `DbContext` class, as shown in the following code block:

```
public class FormsAppDbContext :
AbpDbContext<FormsAppDbContext>
{
    ...

    protected override void OnModelCreating(ModelBuilder
builder)
    {
        base.OnModelCreating(builder);
        // TODO: configure entities...
    }
}
```

When you override the `OnModelCreating` method, always call `base.OnModelCreating()` since ABP also performs default configurations inside that method, necessary to properly use ABP features such as audit logs and data filters. Then, you can use the `builder` object to perform your configurations.

For example, we can configure the mapping for the `Form` class defined in this chapter, as follows:

```
builder.Entity<Form>(b =>
{
    b.ToTable("Forms");
    b.ConfigureByConvention();
    b.Property(x => x.Name)
        .HasMaxLength(100)
        .IsRequired();
    b.HasIndex(x => x.Name);
});
```

Calling the `b.ConfigureByConvention()` method is important here. It configures the base properties of your entity if it is derived from ABP's pre-defined `Entity` or `AggregateRoot` classes. The remaining configuration code is pretty clean and standard, and you can learn all the details from EF Core's documentation.

Here is another example that configures a relation between entities:

```
builder.Entity<Question>(b =>
{
    b.ToTable("FormQuestions");
    b.ConfigureByConvention();
    b.Property(x => x.Title)
        .HasMaxLength(200)
        .IsRequired();
    b.HasOne<Form>()
        .WithMany(x => x.Questions)
        .HasForeignKey(x => x.FormId)
        .IsRequired();
});
```

In this example, we are defining the relation between the Form and Question entities: a form can have many questions, while a question always belongs to a single form.

The configuration we've made ensures that EF Core knows how to read and write entities to the database tables. However, related tables in the database should also be available. You can definitely create a database and the tables inside it manually. Then, in every change to your entities, you manually reflect the related changes in the database schema. However, it is hard to keep your entities and database tables in sync in this way. It is also tedious and error-prone to make them all manual, especially when you have multiple environments (such as development and production).

Fortunately, there is a better way: Code First Migrations. EF's Code First Migrations system provides an efficient way to incrementally update the database schema to keep it in sync with your entity model. We've already used the Code First Migration system in *Chapter 3, Step-By-Step Application Development*. You can refer to that chapter to learn how to add a new database migration and apply it in the database.

Implementing custom repositories

We created an IFormRepository interface in the *Custom repositories* part of the *Working with repositories* section in this chapter. Now, it's time to implement this repository interface using EF Core.

You can implement the repository inside the EF Core integration project of your solution, like this:

```
public class FormRepository :
    EfCoreRepository<FormsAppDbContext, Form, Guid>,
    IFormRepository
{
    public FormRepository(
        IDbContextProvider<FormsAppDbContext>
dbContextProvider)
        : base(dbContextProvider)
    { }


    public async Task<List<Form>> GetListAsync(
        string name, bool includeDrafts = false)
    {
        var dbContext = await GetDbContextAsync();
        var query = dbContext.Forms
            .Where(f => f.Name.Contains(name));
        if (!includeDrafts)
        {
            query = query.Where(f => !f.IsDraft);
        }
        return await query.ToListAsync();
    }
}
```

This class is derived from ABP's `EfCoreRepository` class. In this way, we are inheriting all the standard repository methods. The `EfCoreRepository` class gets three generic parameters: the `DbContext` type, the entity type, and the PK type of the entity class.

`FormRepository` also implements `IFormRepository`, which defines a custom `GetListAsync` method. We get the `DbContext` instance to use all the power of the EF Core API in this method.

Tip about WhereIf

Conditional filtering is a widely used pattern, and ABP provides a nice `WhereIf` extension method that can simplify our code.

We could rewrite the `GetListAsync` method, as shown in the following code block:

```
var dbContext = await GetDbContextAsync();
return await dbContext.Forms
    .Where(f => f.Name.Contains(name))
    .WhereIf(!includeDrafts, f => !f.IsDraft)
    .ToListAsync();
```

Since we have the `DbContext` instance, we can use it to execute **Structured Query Language** (**SQL**) commands or stored procedures. The following method executes a raw SQL command to delete all draft forms:

```
public async Task DeleteAllDraftsAsync()
{
    var dbContext = await GetDbContextAsync();
    await dbContext.Database
        .ExecuteSqlRawAsync("DELETE FROM Forms WHERE
IsDraft = 1");
}
```

> **Executing Stored Procedures and Functions**
>
> You can refer to EF Core's documentation (`https://docs.microsoft.com/en-us/ef/core`) to learn how to execute stored procedures and functions.

Once you implement `IFormRepository`, you can inject and use it instead of `IRepository<Form, Guid>`, as follows:

```
public class FormService : ITransientDependency
{
    private readonly IFormRepository _formRepository;

    public FormService(IFormRepository formRepository)
    {
        _formRepository = formRepository;
    }

    public async Task<List<Form>> GetFormsAsync(string
name)
```

```
    {
        return await _formRepository
            .GetListAsync(name, includeDrafts: true);
    }
}
```

This class uses the custom `GetListAsync` method of `IFormRepository`.

Even if you implement a custom repository class for the `Form` entity, it is still possible to inject and use default generic repositories (for example, `IRepository<Form, Guid>`) for that entity. This is a good feature, especially if you start with generic repositories, then decide to create a custom repository later. You don't have to change your existing code that uses the generic repository.

One potential problem may occur if you override a base method from the `EfCoreRepository` class and customize it in your repository. In this case, the services that use the generic repository reference will continue to use the non-overridden method. To prevent this fragmentation, use the `AddRepository` method while registering your `DbContext` with DI, as follows:

```
context.Services.AddAbpDbContext<FormsAppDbContext>(options =>
{
    options.AddDefaultRepositories();
    options.AddRepository<Form, FormRepository>();
});
```

With this configuration, the `AddRepository` method redirects generic repositories to your custom repository class.

Loading related data

If your entity has navigation properties to other entities or has collections of other entities, then you'll frequently need to access those related entities while working with the main entity. For example, the `Form` entity introduced before has a collection of `Question` entities, and you may need to access the questions while working with a `Form` object.

There are multiple ways to access related entities: **explicit loading**, **lazy loading**, and **eager loading**.

Explicit loading

Repositories provide `EnsurePropertyLoadedAsync` and
`EnsureCollectionLoadedAsync` extension methods to load a navigation property
or sub-collection explicitly.

For example, we can explicitly load the questions of a form, as shown in the following
code block:

```
public async Task<IEnumerable<Question>> GetQuestionsAsync(Form
form)
{
    await _formRepository
        .EnsureCollectionLoadedAsync(form, f =>
f.Questions);
    return form.Questions;
}
```

If we don't use `EnsureCollectionLoadedAsync` here, then the `form.
Questions` collection might be empty. If we are not sure it is filled, we
can use `EnsureCollectionLoadedAsync` to ensure it is loaded. The
`EnsurePropertyLoadedAsync` and `EnsureCollectionLoadedAsync` methods
do nothing if the related property or collection is already loaded, so calling them multiple
times is not a problem for performance.

Lazy loading

Lazy loading is a feature of EF Core that loads related properties and collections when you
first access them. Lazy loading is not enabled by default. If you want to enable it for your
`DbContext`, follow these steps:

1. Install the `Microsoft.EntityFrameworkCore.Proxies` NuGet package in
    your EF Core layer.

2. Use the `UseLazyLoadingProxies` method while configuring
    `AbpDbContextOptions`, as follows:

    ```
    Configure<AbpDbContextOptions>(options =>
    {
        options.PreConfigure<FormsAppDbContext>(opts =>
        {
            opts.DbContextOptions.UseLazyLoadingProxies();
        });
    ```

```
        options.UseSqlServer();
});
```

3. Be sure that the navigation properties and collection properties are virtual in your entities, as shown here:

```
public class Form : BasicAggregateRoot<Guid>
{
    ...
        public virtual ICollection<Question> Questions {
get; set; }
        public virtual ICollection<FormManager> Owners {
get; set; }
}
```

When you enable lazy loading, you don't need to use explicit loading anymore.

Lazy loading is a discussed concept of ORMs. Some developers find it useful and practical, while others suggest not using it in any way. I am drawn to not using it because it has some potential problems, such as these:

- Lazy loading can't use asynchronous programming because there is no way to access a property with the `async`/`await` pattern. So, it blocks the caller thread, which is a bad practice for throughput and scalability.

- You may have a `1+N` loading problem if you forget to eager-load the related data before using a `foreach` loop. `1+N` loading means you query a list of entities from the database with a single database operation (`1`), then perform a loop that accesses a navigation property (or a collection) of these entities. In this case, it lazy-loads the related property for each loop (`N` = count of the queried entities in the first database operation). So, you make a `1+N` database call, which dramatically drops your application performance. You should eager-load the related entities in such cases so that you make a single database in total.

- It makes it hard to predicate and optimize your code since you may not easily see when the related data is loaded from the database.

I suggest going for a more controlled approach and using eager loading wherever possible.

Eager loading

Eager loading is a way of loading related data while first querying the main entity.

Assume that you've created a custom repository method to load the related questions while getting a `Form` object from the database, as shown here:

```
public async Task<Form> GetWithQuestions(Guid formId)
{
    var dbContext = await GetDbContextAsync();
    return await dbContext.Forms
        .Include(f => f.Questions)
        .SingleAsync(f => f.Id == formId);
}
```

If you create such custom repository methods, you can use the full EF Core API. However, if you are working with ABP's repositories and don't want to depend on EF Core in your application layer, you can't use EF Core's `Include` extension method (which is used to eager-load the related data). In this case, you have two options, which are discussed in the next sections.

IRepository.WithDetailsAsync

The `WithDetailsAsync` method of the `IRepository` returns an `IQueryable` instance by including the given properties or collections, as follows:

```
public async Task EagerLoadDemoAsync(Guid formId)
{
    var queryable = await _formRepository
        .WithDetailsAsync(f => f.Questions);
    var query = queryable.Where(f => f.Id == formId);
    var form = await
_asyncExecuter.FirstOrDefaultAsync(query);
    foreach (var question in form.Questions)
    {
        //...
    }
}
```

`WithDetailsAsync(f => f.Questions)` returns `IQueryable<Form>` with questions included, so we can safely loop through the `form.Questions` collection. `IAsyncQueryableExecuter` was explained before, in the *Generic repositories* section of this chapter. The `WithDetailsAsync` method can get more than one expression to include more than one property if you need it. `WithDetailsAsync` can't be used if you need nested includes (the `ThenInclude` extension method in EF Core). In this case, create a custom repository method.

The Aggregate pattern

The Aggregate pattern will be covered in depth in *Chapter 10, DDD – The Domain Layer*. However, to give a brief bit of information, an aggregate is considered a single unit; it is read and saved as a single unit with all sub-collections. That means you always load related questions while loading a form.

ABP supports the aggregate pattern well and allows you to configure eager loading for an entity at a global point. We can write the following configuration inside the `ConfigureServices` method of our module class (in the `EntityFrameworkCore` project in your solution):

```
Configure<AbpEntityOptions>(options =>
{
    options.Entity<Form>(orderOptions =>
    {
        orderOptions.DefaultWithDetailsFunc = query =>
query
            .Include(f => f.Questions)
            .Include(f => f.Owners);
    });
});
```

It is suggested to include all sub-collections. Once you configure the `DefaultWithDetailsFunc` method as shown, then the following will occur:

- Repository methods that return a single entity (such as `GetAsync`) will eager-load related entities by default unless you explicitly disable that behaviour by specifying the `includeDetails` parameter to `false` on the method call.

- Repository methods that return multiple entities (such as `GetListAsync`) will allow the eager loading of related entities, while they will not eager-load by default.

Here are some examples.

Get a single form with sub-collections included like this:

```
var form = await _formRepository.GetAsync(formId);
```

Get a single form without sub-collections like this:

```
var form = await _formRepository.GetAsync(formId,
includeDetails: false);
```

Get a list of forms without sub-collections like this:

```
var forms = await _formRepository.GetListAsync(f => f.Name.
StartsWith("A"));
```

Get a list of forms with sub-collections included like this:

```
var forms = await _formRepository.GetListAsync(f => f.Name.
StartsWith("A"), includeDetails: true);
```

The Aggregate pattern simplifies your application code in most cases, while you can still fine-tune cases where you need performance optimization. Note that navigation properties (to other aggregates) are not used if you truly implement the Aggregate pattern. We will return to this topic again in *Chapter 10, DDD – The Domain Layer*.

We've covered the essentials of using EF Core with ABP Framework. The next section will explain MongoDB integration, the other built-in database provider of ABP Framework.

MongoDB integration

MongoDB is a popular non-relational **document database**, which stores data in JSON-like documents rather than traditional row-/column-based tables.

The ABP CLI provides an option to create new applications using MongoDB, as shown here:

```
abp new FormsApp -d mongodb
```

If you want to check and change the database connection string, you can look at the appsettings.json file of your application.

> **The MongoDB Client Package**
>
> ABP uses the official MongoDB.Driver NuGet package for MongoDB integration.

In the next chapters, you will learn how to work with ABP's `AbpMongoDbContext` class to define `DbContext` objects, perform object-mapping configurations, register `DbContext` objects with the DI system, and implement custom repositories when you want to extend the generic repositories for your entities.

We begin the MongoDB integration by defining a `DbContext` class.

Defining DbContexts

The MongoDB driver package doesn't have a `DbContext` concept like EF Core does. However, ABP introduces the `AbpMongoDbContext` class to provide a standard way to define and configure MongoDB integration. We need to define a class deriving from the `AbpMongoDbContext` base class, as follows:

```
public class FormsAppDbContext : AbpMongoDbContext
{

    [MongoCollection("Forms")]
    public IMongoCollection<Form> Forms =>
Collection<Form();

}
```

The `MongoCollection` attribute sets the collection name on the database side. It is optional and uses the driver's default value if you don't specify it. Defining a collection property on the `FormsAppDbContext` class is required to use the default generic repositories.

Configuring object mappings

While the MongoDB C# driver is not an ORM, it still maps your entities to collections in the database, and you may want to customize the mapping configuration. In this case, override the `CreateModel` method in your `DbContext` class like this:

```
protected override void CreateModel(IMongoModelBuilder builder)
{

    builder.Entity<Form>(b =>
    {

        b.BsonMap.UnmapProperty(f => f.Description);
    });

}
```

In this example, I've configured MongoDB so that it ignores the `Description` property of the `Form` entity while saving and retrieving data. Please refer to the documentation of the `MongoDB.Driver` NuGet package to learn about all configuration options.

Registering DbContext with DI

Once you create and configure your `DbContext` class, it is registered with the DI system in the `ConfigureServices` method of your module class (typically in the MongoDB integration project of your solution). The following code snippet illustrates this:

```
public override void ConfigureServices(
    ServiceConfigurationContext context)
{
    context.Services.AddMongoDbContext<FormsAppDbContext>(
        options =>
            {
                options.AddDefaultRepositories();
            });
}
```

`AddDefaultRepositories()` is used to enable default generic repositories for your entities related to that `DbContext`. You can then inject `IRepository<Form>` into your classes and start using your MongoDB database.

The `AddDefaultRepositories` method enables default repositories only for aggregate root entities (the entity classes derived from the `AggregateRoot` class). Set `includeAllEntities` to `true` to enable default repositories for all entity types. However, it is strongly suggested to apply the Aggregate pattern while working with MongoDB. The Aggregate pattern will be covered in depth in *Chapter 10, DDD – The Domain Layer*.

Default generic repositories are enough in most cases, but you may need to access the MongoDB API or abstract your queries into custom repository methods.

Implementing custom repositories

We created an `IFormRepository` interface in the *Custom repositories* part of the *Working with repositories* section in this chapter. We can implement this repository interface using MongoDB.

You can implement the repository inside the MongoDB integration project of your solution, like this:

```
using System;
using System.Collections.Generic;
using System.Threading.Tasks;
using MongoDB.Driver;
using MongoDB.Driver.Linq;
using Volo.Abp.Domain.Repositories.MongoDB;
using Volo.Abp.MongoDB;
namespace FormsApp
{
    public class FormRepository :
        MongoDbRepository<FormsAppDbContext, Form, Guid>,
        IFormRepository
    {
        public FormRepository(
            IMongoDbContextProvider<FormsAppDbContext>
dbContextProvider)
            : base(dbContextProvider)
        { }
        // TODO: implement the GetListAsync method
    }
}
```

The `FormRepository` class is derived from ABP's `MongoDbRepository` class. In this way, we are inheriting all the standard repository methods. The `MongoDbRepository` class gets three generic parameters: the `DbContext` type, the entity type, and the PK type of the entity class.

The `FormRepository` class should implement the `GetListAsync` method defined by the `IFormRepository` interface, as follows:

```
public async Task<List<Form>> GetListAsync(
    string name, bool includeDrafts = false)
{
    var queryable = await GetMongoQueryableAsync();
    var query = queryable.Where(f =>
f.Name.Contains(name));
```

```
    if (!includeDrafts)
    {
        query = queryable.Where(f => !f.IsDraft);
    }

    return await query.ToListAsync();
}
```

I've used the LINQ API of the MongoDB driver in this example, but you can use alternative APIs by obtaining the `IMongoCollection` object, as illustrated in the following code snippet:

```
IMongoCollection<Form> formsCollection = await
GetCollectionAsync();
```

Now, you can inject `IFormRepository` instead of the generic `IRepository<Form, Guid>` repository into your services and use all the standard and custom repository methods.

Even if you implement a custom repository class for the `Form` entity, it is still possible to inject and use default generic repositories (such as `IRepository<Form, Guid>`) for that entity. If you implement a custom repository, it is suggested to use the `AddRepository` method on the `DbContext` registration code, as illustrated in the following code snippet:

```
context.Services.AddMongoDbContext<FormsAppDbContext>(options
=>
{
    options.AddDefaultRepositories();
    options.AddRepository<Form, FormRepository>();
});
```

In this way, generic default repositories will be redirected to your custom repository class. If you override a base method in your custom repository, they will also use your overload instead of the base method.

We've learned how to use EF Core and MongoDB as the database provider. In the next section, we will understand the UoW system, making it possible to connect these databases and apply transactions.

Understanding the UoW system

UoW is the main system that ABP uses to initiate, manage, and dispose of database connections and transactions. The UoW system is designed with the **Ambient Context pattern**. That means when we create a new UoW, it creates a scoped context that is participated by all the database operations performed in the current scope by sharing the same context and is considered a single transaction boundary. All the operations done in a UoW are committed (on success) or rolled back (on exception) together.

While you can manually create UoW scopes and control the transaction properties, most of the time, it works seamlessly just as you desire. However, it provides some options if you change the default behavior.

> **UoW and Database Operations**
>
> All database operations must be performed in a UoW scope since UoW is the way to manage database connections and transactions in ABP Framework. Otherwise, you get an exception indicating that.

In the next sections, you will gain an understanding of how the UoW system works and customize it by configuring the options. I will also explain how to manually control the UoW system when the conventional system doesn't work for your use case.

Configuring UoW options

With the default setup, in an ASP.NET Core application, an HTTP request is considered as the UoW scope. ABP starts a UoW at the beginning of the request and saves changes to the database if the request successfully finishes. It rolls back the UoW if the request fails because of an exception.

ABP determines database transaction usage based on the HTTP request type. HTTP GET requests don't create a database transaction. UoW works anyway but doesn't use a database transaction in this case. All other HTTP request types (POST, PUT, DELETE, and others) use a database transaction if you haven't configured them otherwise.

> **HTTP GET Requests and Transactions**
>
> It is a best practice not to make database changes in GET requests. If you make multiple write operations in a GET request and somehow your request fails, your database state could be left in an inconsistent state because ABP doesn't create a database transaction for GET requests. In this case, either enable transactions for GET requests using AbpUnitOfWorkDefaultOptions or manually control the UoW, as described in the next section.

Use `AbpUnitOfWorkDefaultOptions` in the `ConfigureServices` method of your module (in the database integration project) if you want to change the UoW options, as follows:

```
public override void ConfigureServices(
    ServiceConfigurationContext context)
{
    Configure<AbpUnitOfWorkDefaultOptions>(options =>
    {
        options.TransactionBehavior =
UnitOfWorkTransactionBehavior.Enabled;
        options.Timeout = 300000; // 5 minutes
        options.IsolationLevel =
IsolationLevel.Serializable;
    });
}
```

`TransactionBehavior` can take the following three values:

- `Auto` (default): Automatically determines using database transactions (transactions are enabled for non-GET HTTP requests)
- `Enabled`: Always uses a database transaction, even for HTTP GET requests
- `Disabled`: Never uses a database transaction

The `Auto` behavior is the default value and is suggested for most applications. `IsolationLevel` is only valid for relational databases. ABP uses the default value of the underlying provider if you don't specify it. Finally, the `Timeout` option allows you to set a default timeout value for transactions as milliseconds. If a UoW operation doesn't complete in the given timeout value, a timeout exception is thrown.

In this section, we've learned how to configure the default options across all UoWs. It is also possible to configure these values for an individual UoW if you manually control it.

Manually controlling the UoW

For web applications, you rarely need to control the UoW system manually. However, for background workers or non-web applications, you may need to create UoW scopes yourself. You may also need to control the UoW system to create inner transaction scopes.

One way to create a UoW scope is to use the `[UnitOfWork]` attribute on your method, like this:

```
[UnitOfWork(isTransactional: true)]
public async Task DoItAsync()
{
    await _formRepository.InsertAsync(new Form() { ... });
    await _formRepository.InsertAsync(new Form() { ... });
}
```

The UoW system uses the Ambient Context pattern. If a surrounding UoW is already in place, your `UnitOfWork` attribute is ignored and your method participates in the surrounding UoW. Otherwise, ABP starts a new transactional UoW just before entering the `DoItAsync` method and commits the transaction if it doesn't throw an exception. The transaction is rolled back if that method throws an exception.

If you want to fine-control the UoW system, you can inject and use the `IUnitOfWorkManager` service, as shown in the following code block:

```
public async Task DoItAsync()
{
    using (var uow = _unitOfWorkManager.Begin(
        requiresNew: true,
        isTransactional: true,
        timeout: 15000))
    {
        await _formRepository.InsertAsync(new Form() { });
        await _formRepository.InsertAsync(new Form() { });
        await uow.CompleteAsync();
    }
}
```

In this example, we are starting a new transactional UoW scope with 15 seconds as the `timeout` parameter's value. With this usage (`requiresNew: true`), ABP always starts a new UoW even if there is a surrounding UoW. Always call the `uow.CompleteAsync()` method if everything goes right. You can use the `uow.RollbackAsync()` method if you want to roll back the current transaction.

As mentioned before, UoW uses an ambient scope. You can access the current UoW anywhere in this scope, using the IUnitOfWorkManager.Current property. It can be null if there is no ongoing UoW.

The following code snippet uses the SaveChangesAsync method with the IUnitOfWorkManager.Current property:

```
await _unitOfWorkManager.Current.SaveChangesAsync();
```

We've saved all pending changes to the database. However, if that's a transactional UoW, these changes are also rolled back if you roll back the UoW or throw any exception in the UoW scope.

Summary

In this chapter, we've learned how to work with databases using ABP Framework. ABP standardizes to define entities by providing base classes. It also helps to automatically track change times and the users changing entities, when you derive from audited entity classes.

The repository system provides the fundamental functionalities to read and write entities. You can use LINQ over the repositories for advanced querying possibilities. Also, you can create custom repository classes to work with the underlying data provider directly, hide complex queries behind simple repository interfaces, call stored procedures, and so on.

ABP is database-agnostic, but it provides integration packages with EF Core and MongoDB out of the box. ABP application startup templates come with one of these providers, whichever you prefer.

EF Core is the de facto ORM for the .NET platform, and ABP supports EF Core as a first-class citizen. The application startup template is fine-tuned to configure your mappings and manage your database schema migrations while supporting a modular application structure.

Finally, the UoW system provides a seamless way to manage database connections and transactions for us. It keeps the application code clean by automating these repeating tasks for us.

Data access is a core requirement for any business application, and it is essential to understand the details of it. The next chapter will continue with the cross-cutting concerns required for every application, such as authorization, validation, and exception handling.

7
Exploring Cross-Cutting Concerns

Cross-cutting concerns such as authorization, validation, exception handling, and logging are fundamental parts of any serious system. They are essential to make your system secure and operate well.

One problem with implementing cross-cutting concerns is that you should implement these concerns everywhere in your application, which leads to a repetitive code base. Also, one missing authorization or validation check may explode your entire system.

One of ABP Framework's main goals is to help you apply the **Don't Repeat Yourself (DRY)** principle! ASP.NET Core already provides a good infrastructure for some cross-cutting concerns, but ABP takes it further to automate or make them much easier for you.

This chapter explores ABP's infrastructure for the following cross-cutting concerns:

- Working with authorization and permission systems
- Validating user inputs
- Exception handling

Technical requirements

If you want to follow along and try the examples, you need to install an **integrated development environment** (**IDE**)/editor (for example, Visual Studio) to build the ASP.NET Core projects.

You can download the code examples from the following GitHub repository: `https://github.com/PacktPublishing/Mastering-ABP-Framework`.

This chapter also references the *EventHub* project for some code examples. That project was introduced in *Chapter 4, Understanding the Reference Solution*, and you can access its source code from the following GitHub repository: `https://github.com/volosoft/eventhub`.

Working with authorization and permission systems

Authentication and **authorization** are two major concepts in software security. Authentication is the process of identifying the current user. On the other hand, authorization is used to allow or prohibit a user from performing a specific action in the application.

ASP.NET Core's authorization system provides an advanced and flexible way to authorize the current user. ABP Framework's authorization infrastructure is 100% compatible with ASP.NET Core's authorization system and extends it by introducing the permission system. ABP allows permissions to be easily granted to roles and users. It allows the same permissions to be checked on the client side too.

I will explain the authorization system as a mix of ASP.NET Core's and ABP's infrastructure by indicating which part is added by ABP Framework. Let's begin with the simplest authorization check.

Simple authorization

In the simplest case, you may want to allow a certain operation only for those logged in to the application. The `[Authorize]` attribute, without any parameters, only checks whether the current user has been authenticated (logged in).

See the following **model-view-controller (MVC)** example:

```
public class ProductController : Controller
{
    public async Task<List<ProductDto>> GetListAsync()
    {
    }
    [Authorize]
    public async Task CreateAsync(ProductCreationDto input)
    {
    }
    [Authorize]
    public async Task DeleteAsync(Guid id)
    {
    }
}
```

In this example, the `CreateAsync` and `DeleteAsync` actions are only usable by authenticated users. Suppose an anonymous user (a user that has not logged in to the application, so we couldn't identify them) tries to execute these actions. In that case, ASP.NET Core returns an authorization error response to the client. However, the `GetListAsync` method is available to everyone, even to anonymous users.

The `[Authorize]` attribute can be used at the controller class level to authorize all the actions inside that controller. In that case, we can use the `[AllowAnonymous]` attribute to allow a specific action to anonymous users. So, we could rewrite the same example, as shown in the following code block:

```
[Authorize]
public class ProductController : Controller
{
    [AllowAnonymous]
    public async Task<List<ProductDto>> GetListAsync()
    {
    }

    public async Task CreateAsync(ProductCreationDto input)
    {
    }
```

```
    public async Task DeleteAsync(Guid id)
    {
    }
}
```

Here, I used the `[Authorize]` attribute on top of the class and added `[AllowAnonymous]` to the `GetListAsync` method. This makes it possible to also consume that particular action for users who haven't logged in to the application.

While the parameterless `[Authorize]` attribute has some use cases, you generally want to define specific permissions (or policies) in your application so that all authenticated users don't have the same privileges.

Using the permission system

The most important authorization extension of ABP Framework for ASP.NET Core is the permission system. A permission is a simple policy that is granted or prohibited for a particular user or role. It is then associated with a particular functionality of your application and is checked when users try to use that functionality. If the current user has the related permission granted, then the user can use the application functionality. Otherwise, the user cannot use that functionality.

ABP provides all the functionality to define, grant, and check permissions in your application.

Defining permissions

We should define permissions before using them. To define permissions, create a class that inherits from the `PermissionDefinitionProvider` class. When you create a new ABP solution, an empty permission definition provider class comes in the `Application.Contracts` project of the solution. See the following example:

```
public class ProductManagementPermissionDefinitionProvider
    : PermissionDefinitionProvider
{
    public override void Define(
        IPermissionDefinitionContext context)
    {
        var myGroup = context.AddGroup(
            "ProductManagement");
```

```
            myGroup.AddPermission(
                "ProductManagement.ProductCreation");
            myGroup.AddPermission(
                "ProductManagement.ProductDeletion");
        }
    }
```

ABP Framework calls the `Define` method on application startup. In this example, I've created a permission group, named `ProductManagement`, and defined two permissions inside it. Groups are used to group permissions on the **user interface** (**UI**), and generally, every module defines its permission group. Group and permission names are arbitrary `string` values (it is suggested to define `const` fields instead of using magic strings).

That was a minimal configuration. You can also specify display names as localizable strings for the group, and permission names to show them in a user-friendly way on the UI. The following code block uses the localization system to specify the display names while defining the group and the permissions:

```
public class ProductManagementPermissionDefinitionProvider
    : PermissionDefinitionProvider
{
    public override void Define(
        IPermissionDefinitionContext context)
    {
        var myGroup = context.AddGroup(
            «ProductManagement»,
            L("ProductManagement"));
        myGroup.AddPermission(
            "ProductManagement.ProductCreation",
            L("ProductCreation"));
        myGroup.AddPermission(
            "ProductManagement.ProductDeletion",
            L("ProductDeletion"));
    }

    private static LocalizableString L(string name)
    {
        return LocalizableString
```

```
            .Create<ProductManagementResource>(name);
    }
}
```

I've defined an `L` method to simplify the localization. The localization system will be covered in *Chapter 8, Using the Features and Services of ABP*.

> **Permission Definitions in Multi-Tenant Applications**
>
> For multi-tenant applications, you can specify the `multiTenancySide` parameter for the `AddPermission` method, to define host-only or tenant-only permissions. We will return to this topic in *Chapter 16, Implementing Multi-Tenancy*.

Once you define a permission, it becomes available on the permission management dialog after the next application startup.

Managing permissions

A permission can be granted for a user or role by default. For example, assume that you have created a manager role and want to grant the product permissions for that role. When you run the application, navigate to the **Administration | Identity Management | Roles** page. Then create the `manager` role if you haven't created it before; to do so, click on the **Actions** button and select the **Permissions** action, as shown in *Figure 7.1*:

Figure 7.1 – Selecting the Permissions action on the Role Management page

Clicking on the **Permissions** action opens a modal dialog to manage the permissions of the selected role, as shown here:

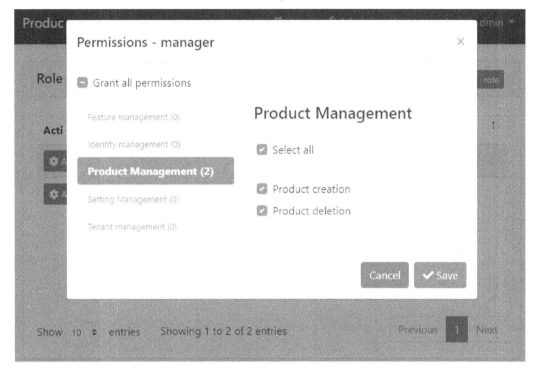

Figure 7.2 – Permission Management modal

In *Figure 7.2*, you see the permission groups on the left-hand side, while the permissions in this group are available on the right-hand side. The permission group and the permissions we've defined are available in this dialog box without any additional effort.

All users with the manager role inherit the permissions of that role. Users can have multiple roles, and they inherit a union of all permissions of all the assigned roles. You can also grant permissions directly to the users on the user management page for more flexibility.

We've defined permissions and assigned them to roles. The next step is to check whether the current user has the requested permissions.

Checking permissions

You can check a permission either declaratively, using the [Authorize] attribute, or programmatically, using IAuthorizationService.

We can rewrite the `ProductController` class (which was introduced in the *Simple authorization* section) to request the product creation and deletion permissions on specific actions, as follows:

```
public class ProductController : Controller
{
    public async Task<List<ProductDto>> GetListAsync()
    {
    }
    [Authorize("ProductManagement.ProductCreation")]
    public async Task CreateAsync(ProductCreationDto input)
    {
    }
    [Authorize("ProductManagement.ProductDeletion")]
    public async Task DeleteAsync(Guid id)
    {
    }
}
```

The `[Authorize]` attribute, with this usage, takes a string parameter as a policy name. ABP defines permissions as automatic policies, so you can use a permission name wherever you need to specify a policy name.

Declarative authorization is simple to use and recommended wherever possible. However, it is limited when you want to check permissions or perform logic for unauthorized cases conditionally. For such cases, you can inject and use `IAuthorizationService`, as shown in the following example:

```
public class ProductController : Controller
{
    private readonly IAuthorizationService
        _authorizationService;

    public ProductController(
        IAuthorizationService authorizationService)
    {
        _authorizationService = authorizationService;
    }
```

```
public async Task CreateAsync(ProductCreationDto input)
{
    if (await _authorizationService.IsGrantedAsync(
        "ProductManagement.ProductCreation"))
    {
        // TODO: Create the product
    }
    else
    {
        // TODO: Handle unauthorized case
    }
}
}
```

The IsGrantedAsync method checks the given permission and returns true if the current user (or a role of the user) has been granted the current permission. This is useful if you have custom logic for an unauthorized case. However, if you want to simply check the permission and throw an exception for unauthorized cases, the CheckAsync method is more practical:

```
public async Task CreateAsync(ProductCreationDto input)
{
    await _authorizationService
        .CheckAsync("ProductManagement.ProductCreation");
    //TODO: Create the product
}
```

The CheckAsync method throws an AbpAuthorizationException exception if the user doesn't have permission to carry out that operation, which is handled by ABP Framework to return a proper **HyperText Transfer Protocol** (**HTTP**) response to the client (this will be discussed in the *Exception handling* section of this chapter). The IsGrantedAsync and CheckAsync methods are useful extension methods defined by ABP Framework.

Tip: Inherit from AbpController

It is suggested to derive your controller classes from the `AbpController` class instead of the standard `Controller` class. This extends the standard `Controller` class and defines some useful base properties. For example, it has the `AuthorizationService` property (of the `IAuthorizationService` type), which you can directly use instead of manually injecting the `IAuthorizationService` interface.

Checking permissions on the server is a common approach. However, you may also need to check permissions on the client side.

Using permissions on the client side

ABP exposes a standard HTTP API with a URL of `/api/abp/application-configuration`, which returns JSON data containing localization texts, settings, permissions, and more. Then, the client application can consume that API to check permissions or perform localization on the client side.

Different client types may provide different services to check permissions. For example, in an MVC/Razor Pages application, you can use the `abp.auth` JavaScript API to check a permission, as illustrated here:

```
abp.auth.isGranted('ProductManagement.ProductCreation');
```

This is a global function that returns `true` if the current user has the given permission. Otherwise, it returns `false`.

In a Blazor application, you can reuse the same `[Authorize]` attribute and `IAuthorizationService`.

We will return to client-side permission checking in *Part 4, User Interface and API Development*.

Child permissions

In a complex application, you may need to create some child permissions that depend on their parent permissions. The child permissions are meaningful only if the parent permission has been granted. See *Figure 7.3*:

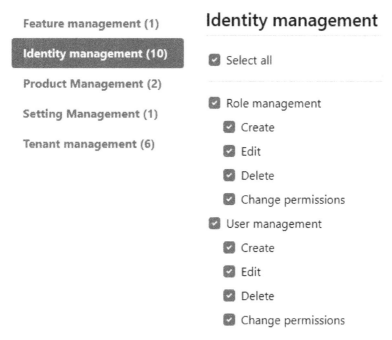

Figure 7.3 – Parent-child permissions

In *Figure 7.3*, the **Role management** permission has some child permissions such as **Create**, **Edit**, and **Delete**. The **Role management** permission is used to allow a user to enter the **Role Management** page. If the user cannot enter the page, then there is no point in granting the role creation permission, since it is practically impossible to create a new role without entering that page.

In the permission definition class, the AddPermission method returns the created permission so that you can assign it to a variable and use the AddChild method to create a child permission, as shown in the following code block:

```
public override void Define(IpermissionDefinitionContext
                            context)
{
    var myGroup = context.AddGroup(
        "ProductManagement",
        L("ProductManagement"));
    var parent = myGroup.AddPermission(
        "MyParentPermission");
    parent.AddChild("MyChildPermission");
}
```

In this example, we've created a permission named `MyParentPermission`, then created another permission named `MyChildPermission` as a child permission.

Child permissions can also have child permissions. You can assign the return value of the `parent.AddChild` method to a variable and call its `AddChild` method.

Defining and using permissions is an easy yet powerful way to authorize an application via simple on/off-style policies. However, ASP.NET Core allows the creation of complete custom logic to define policies.

Policy-based authorization

The ASP.NET Core **policy-based authorization** system allows you to authorize certain operations in your application, just as with permissions, but this time, with your custom logic expressed with code. Actually, a permission is a simplified and automated policy provided by ABP Framework.

Assume that you want to authorize a *product creation* operation with your custom code. You first need to define a requirement that you will check later (we can define these classes in the application layer of the solution, while there is no strict rule). The code is illustrated in the following snippet:

```
public class ProductCreationRequirement :
    IAuthorizationRequirement
{ }
```

`ProductCreationRequirement` is an empty class that just implements the `IAuthorizationRequirement` marker interface. Then, you should define an authorization handler for that requirement, as follows:

```
public class ProductCreationRequirementHandler
    : AuthorizationHandler<ProductCreationRequirement>
{
    protected override Task HandleRequirementAsync(
        AuthorizationHandlerContext context,
        ProductCreationRequirement requirement)
    {
        if (context.User.HasClaim(c => c.Type ==
            "productManager"))
        {
            context.Succeed(requirement);
```

```
        }

        return Task.CompletedTask;
    }
}
```

The handler class must be derived from `AuthorizationHandler<T>`, where T is the type of your requirement class. In this example, I simply checked whether the current user has the `productManager` claim, which is my custom claim (a claim is a simple named value stored in the authentication ticket). You can build your custom logic. All you're going to do is call `context.Succeed` if you want to allow the current user to have the requirement.

Once you define a requirement and handler, you need to register them in the `ConfigureServices` method of your module class, like this:

```
public override void ConfigureServices(
    ServiceConfigurationContext context)
{
    Configure<AuthorizationOptions>(options =>
    {
        options.AddPolicy(
            "ProductManagement.ProductCreation",
            policy => policy.Requirements.Add(
                new ProductCreationRequirement()
            )
        );
    });
    context.Services.AddSingleton<IAuthorizationHandler,
        ProductCreationRequirementHandler>();
}
```

I've used `AuthorizationOptions` to define a policy named `ProductManagement.ProductCreation` with the `ProductCreationRequirement` requirement. Then, I've registered `ProductCreationRequirementHandler` as a singleton service.

Now, suppose I use the `[Authorize("ProductManagement.ProductCreation")]` attribute on a controller or action or use `IAuthorizationService` to check the policy. In that case, my custom authorization handler logic works to allow me to take complete control of the policy-check logic.

> **Permissions versus Custom Policies**
>
> Once you implement a custom policy, you cannot use the permission management dialog to grant permission to users and roles because it is not a simple on/off permission that you can enable/disable. However, a client-side policy check still works, since ABP is well integrated into ASP.NET Core's policy system.

As you can see, ABP's permission system is much easier and more powerful if you just need on/off-style policies, while custom policies allow you to dynamically check policies with your custom logic.

> **Resource-Based Authorization**
>
> ASP.NET Core's authorization system has more features than covered here. Resource-based authorization is one feature that allows you to control policies based on objects (such as entities). For example, you can control access to delete a specific product, rather than having a common deleting permission for all products. ABP is 100% compatible with the ASP.NET Core authorization system, so I suggest you check ASP.NET Core's documentation to learn more about authorization: `https://docs.microsoft.com/en-us/aspnet/core/security/authorization`.

Up to now, we've seen the usage of the `[Authorize]` attribute on MVC controllers. However, this attribute and `IAuthorizationService` are not limited to controllers.

Authorizations outside of controllers

ASP.NET Core allows you to use the `[Authorize]` attribute and `IAuthorizationService` for Razor Pages, Razor components, and some other points in the web layer. You can refer to ASP.NET Core's documentation to learn about these standard usages: `https://docs.microsoft.com/en-us/aspnet/core/security/authorization`.

ABP Framework takes it one step further and allows using the [Authorize] attribute for the application service classes and methods without depending on the web layer, even in a non-web application. So, this usage is completely valid, as illustrated here:

```
public class ProductAppService
    : ApplicationService, IProductAppService
{
    [Authorize("ProductManagement.ProductCreation")]
    public Task CreateAsync(ProductCreationDto input)
    {
        // TODO
    }
}
```

The CreateAsync method can only be executed if the current user has the ProductManagement.ProductCreation permission/policy. Actually, [Authorize] is usable in any class that is registered for **dependency injection (DI)**. However, since authorization is considered an application layer aspect, it is recommended that authorization be used at the application layer and not at the domain layer.

> **Dynamic Proxying/Interceptors**
>
> ABP uses dynamic proxying using interceptors to accomplish the authorization check on method calls. If you inject a service via a class reference (rather than an interface reference), the dynamic proxying system uses the dynamic inheritance technique. In this case, your method must be defined with the virtual keyword to allow the dynamic proxying system to override it and perform the authorization check.

The authorization system guarantees that only authorized users consume your services. It is one of the systems that you need to use to secure your application, while the other one is input validation.

Validating user inputs

Validation ensures your data security and consistency and helps your application to operate properly. Validation is a wide topic, and there are some common levels of validation, as outlined here:

- **Client-side validation** is used to pre-validate the user input before sending data to the server. It is important for the **user experience** (**UX**), and you should always implement it wherever possible. However, it cannot guarantee security—even an inexperienced hacker can bypass it. For example, checking whether a required textbox field is empty is a type of client-side validation. We will cover client-side validation in *Part 4, User Interface and API Development*.

- **Server-side validation** is performed by the server to prevent incomplete, badly formatted, or malicious requests. It provides some level of security for your application and is generally performed when you first touch the data sent by the client. For example, checking a required input field is empty on the server side is an example of this type of validation.

- **Business validation** is also performed in the server; it implements your business rules and keeps your business data consistent. It is done at every level of your business code. For example, checking a user's balance before a money transfer is a kind of business validation. We will cover business validation in *Chapter 10, DDD – The Domain Layer*.

> **About the ASP.NET Core Validation System**
>
> ASP.NET Core provides many options for input validation. This book covers the basics by focusing on the features added by ABP Framework. See ASP.NET Core's documentation for all the validation possibilities: `https://docs.microsoft.com/en-us/aspnet/core/mvc/models/validation`.

This section focuses on server-side validation and shows how to perform input validation in different ways. It also explores ways to control the validation process and deal with validation exceptions.

Let's start with the easiest way to perform validation—using data annotation attributes.

Using data annotation attributes

Using data annotation attributes is the simplest way to perform a formal validation for the user input. See the following application service method:

```
public class ProductAppService
    : ApplicationService, IProductAppService
{
    public Task CreateAsync(ProductCreationDto input)
    {
        // TODO
    }
}
```

`ProductAppService` is an application service, and the application service inputs are automatically validated in ABP Framework, just as with controllers in the ASP.NET Core MVC framework. The `ProductAppService` service takes an input parameter, as shown in the following code block:

```
public class ProductCreationDto
{
    [Required]
    [StringLength(100)]
    public string Name { get; set; }

    [Range(0, 999.99)]
    public decimal Price { get; set; }

    [Url]
    public string PictureUrl { get; set; }

    public bool IsDraft { get; set; }
}
```

`ProductCreationDto` has three properties decorated with validation attributes. There are many built-in validation attributes of ASP.NET Core, including the following:

- `[Required]`: Validates that the property is not null

- `[StringLength]`: Validates a maximum (and optionally minimum) length for a string property

- [Range]: Validates that the property value is in the specified range
- [Url]: Validates that the property value has a proper URL format
- [RegularExpression]: Allows the specifying of a custom **regular expression** (**regex**) to validate the property value
- [EmailAddress]: Validates that the property has a properly formatted email address value

ASP.NET Core also allows you to define custom validation attributes by inheriting them from the ValidationAttribute class and overriding the IsValid method.

Data annotation attributes are very easy to use and are suggested to perform formal validation for your **data transfer objects** (**DTOs**) and models. However, they are limited when you need to perform custom code logic to validate the input.

Custom validation with the IValidatableObject interface

A model or DTO object can implement the IValidatableObject interface to perform validation using a custom code block. See the following example:

```
public class ProductCreationDto : IValidatableObject
{
    ...
    [Url]
    public string PictureUrl { get; set; }
    public bool IsDraft { get; set; }
    public IEnumerable<ValidationResult> Validate(
        ValidationContext context)
    {
        if (IsDraft == false &&
            string.IsNullOrEmpty(PictureUrl))
        {
            yield return new ValidationResult(
                "Picture must be provided to publish a
                product",
                new []{ nameof(PictureUrl) }
            );
        }
    }
```

```
        }
    }
```

In this example, ProductCreationDto has a custom rule: a profile picture is required if IsDraft is false. So, we are checking the condition and adding a validation error in this case.

If you need to resolve a service from the DI system, you can use the context. GetRequiredService method. For example, if we want to localize the error message, we can rewrite the Validate method, as shown in the following code block:

```
public IEnumerable<ValidationResult> Validate(
    ValidationContext context)
{
    if (IsDraft == false &&
        string.IsNullOrEmpty(PictureUrl))
    {
        var localizer = context.GetRequiredService
            <IStringLocalizer<ProductManagementResource>
            >();

        yield return new ValidationResult(
            localizer["PictureIsMissingErrorMessage"],
            new []{ nameof(PictureUrl) }
        );
    }
}
```

Here, we resolve an IStringLocalizer<ProductManagementResource> instance from the DI and use it to return a localized error message to the client. We will cover the localization system in *Chapter 8, Using the Features and Services of ABP.*

Formal Validation versus Business Validation

As a best practice, implement only formal validation (such as if a DTO property was not filled or not formatted as expected) in the DTO/model classes and use only the data already available on the DTO/model class. Implement your business validation logic inside application or domain layer services. For example, if you want to check whether a given product name already exists in the database, do not try to implement this logic in the Validate method.

Using either the validation attributes or custom validation logic, ABP Framework handles the validation result and throws an exception prior to the execution of your method.

Understanding the validation exception

If the user input is not valid, ABP Framework automatically throws an exception of the `AbpValidationException` type. The exception is thrown in the following situations:

- The input object is `null`, so you don't need to check whether it is `null`.

- The input object is invalid in any way, so you don't have to check `Model.IsValid` in your API controllers.

ABP doesn't call your service method (or controller action) in these cases. If your method is being executed, you can be sure that the input is not null and is valid.

If you perform additional validation inside your services and want to throw a validation-related exception, you can also throw `AbpValidationException`, as illustrated in the following code snippet:

```
public async Task CreateAsync(ProductCreationDto input)
{
    if (await HasExistingProductAsync(input.Name))
    {
        throw new AbpValidationException(
            new List<ValidationResult>
            {
                new ValidationResult(
                    "Product name is already in use!",
                    new[] {nameof(input.Name)}
                )
            }
        );
    }
}
```

Here, we are assuming that `HasExistingProductAsync` returns `true` if there is a product with the given name. In this case, we are throwing `AbpValidationException` by specifying the validation error(s). `ValidationResult` represents a validation error; its first constructor parameter is the validation error message, and the second parameter (optional) is the name of the DTO properties that caused the validation error.

Once you or the ABP validation system throws an `AbpValidationException` exception, the ABP exception-handling system catches and handles it properly, as we will see in the next section.

The ABP validation system works just as you want most of the time, but sometimes, you may need to bypass it and apply your custom logic.

Disabling the validation

It is possible to bypass the ABP validation system at a method or class level using the `[DisableValidation]` attribute, as in the following example:

```
[DisableValidation]
public async Task CreateAsync(ProductCreationDto input)
{
}
```

In this example, the `CreateAsync` method is decorated with the `[DisableValidation]` attribute, so ABP doesn't perform any automatic validation for the `input` object.

If you use the `[DisableValidation]` attribute for a class, then the validation is disabled for all the methods. In this case, you can use the `[EnableValidation]` attribute for a method to enable validation only for that particular method.

When you disable auto-validation for a method, you can still perform your custom validation logic and throw `AbpValidationException`, as explained in the previous section.

Validation in other types

ASP.NET Core performs validation for controller actions and Razor Page handlers. ABP, in addition to ASP.NET Core, performs validation for application service methods by default.

Beyond the default behavior, ABP allows you to enable the auto-validation feature for any kind of class in your application. All you need to do is to implement the IValidationEnabled marker interface, as shown in the following example:

```
public class SomeServiceWithValidation
    : IValidationEnabled, ITransientDependency
{
    ...
}
```

Then, ABP auto-validates all the inputs for this class, using the validation system explained in this chapter.

> **Dynamic Proxying/Interceptors**
>
> ABP uses dynamic proxying using interceptors to accomplish validation on method calls. If you inject a service via a class reference (rather than an interface reference), the dynamic proxying system uses the dynamic inheritance technique. In this case, your method must be defined with the virtual keyword to allow the dynamic proxying system to override it and perform the validation.

Up to now, we've explained the ABP validation system that is directly compatible with ASP.NET Core's validation infrastructure. The next section introduces FluentValidation library integration, which allows you to separate the validation logic from the validated object.

Integrating the FluentValidation library

The built-in validation system is enough for most cases, and it is easy to use to define formal validation rules. I personally don't see any problem with it and find it practical to embed the data validation logic inside DTO/model classes. However, some developers think that the validation logic inside DTO/model classes is a bad practice, even when it is only a formal validation. In this case, ABP provides an integration package with the popular FluentValidation library, which decouples the validation logic from the DTO/model class and provides more powerful features compared to the standard data annotation approach.

If you want to use the `FluentValidation` library, you first need to install it into your project. You can use the `add-package` command of the **ABP Command-Line Interface (ABP CLI)** to install it for a project easily, as follows:

```
abp add-package Volo.Abp.FluentValidation
```

Once you install the package, you can create your validator classes and set your validation rules, as shown in the following code block:

```
public class ProductCreationDtoValidator
    : AbstractValidator<ProductCreationDto>
{
    public ProductCreationDtoValidator()
    {
        RuleFor(x => x.Name).NotEmpty().MaximumLength(100);
        RuleFor(x => x.Price).ExclusiveBetween(0, 1000);
        //...
    }
}
```

Please refer to the `FluentValidation` documentation to learn how to define advanced validation rules: `https://fluentvalidation.net`.

ABP automatically discovers the validator classes and integrates them into the validation process. That means you can even mix the standard validation logic with the `FluentValidation` validator classes.

Authorization and validation exceptions are well-defined exception types, and ABP automatically handles them. The next section explores the ABP exception-handling system and explains how to deal with different kinds of exceptions.

Exception handling

One of the most important quality indicators of an application is how it responds to errors and exceptional cases. A good application should handle errors, return a proper response to the client, and gracefully inform the user about the problem.

In a typical web application, we should care about exceptions in every client request, which makes it a repetitive and tedious task for developers.

ABP Framework completely automates error handling in every aspect of your application. Most of the time, you don't need to write any `try-catch` statement in your application code, as it does the following:

- Handles all exceptions, logs them, and returns a standard-formatted error response to the client for an API request or shows a standard error page for a server-rendered page

- Hides internal infrastructure errors while allowing you to return user-friendly, localized error messages when you need them

- Understands standard exceptions such as validation and authorization exceptions and sends a proper HTTP status code to the client

- Handles all errors on the client and shows a meaningful message to the end user

While ABP takes care of exceptions, you can throw exceptions to return user-friendly messages or business-specific error codes to the client.

User-friendly exceptions

ABP provides some predefined exception classes to customize the error-handling behavior. One of these is the `UserFriendlyException` class.

First, to understand the need for the `UserFriendlyException` class, see what happens if an arbitrary exception is thrown from a server-side API. The following method throws an exception with a custom message:

```
Public async Task ExampleAsync()
{
    throw new Exception("my error message...");
}
```

Assume that a browser client calls that method via an AJAX request. It will show the following error message to the end user:

An internal error occurred during
your request!

Figure 7.4 – The default error message

As you see in *Figure 7.4*, ABP shows a standard error message about an internal problem. The actual error message is written to the logging system. The server returns an HTTP 500 status code to the client for such generic errors.

That is good behavior because it is not useful to show a raw exception message to an end user. It can even be dangerous, since it may include some sensitive information about your internal systems, such as database table names and fields.

However, you may want to return a user-friendly, informative message to the end user for some specific cases. For such cases, you can throw a `UserFriendlyException` exception, as shown in the following code block:

```
public async Task ExampleAsync()
{
    throw new UserFriendlyException(
        "This message is available to the user!");
}
```

ABP, at this time, doesn't hide the error message, as we can see here:

This message is available to the
user!

Figure 7.5 – Custom error message

The UserFriendlyException class is not unique. Any exception class that inherits from the UserFriendlyException class or directly implements the IUserFriendlyException interface can be used to return user-friendly exception messages. ABP returns an HTTP 403 (forbidden) status code to the client when you throw a user-friendly exception. See the *Controlling the HTTP status code* section of this chapter for all HTTP status code mappings.

In a multilingual application, you will probably want to return a localized message. Use the localization system in this case, which will be introduced in *Chapter 8*, *Using the Features and Services of ABP*.

UserFriendlyException is a special type of business exception where you directly return a message to the user.

Business exceptions

You will have some business rules in a business application, and you need to throw exceptions when the requested operation is not appropriate to execute in the current conditions based on these rules. Business exceptions in ABP are special kinds of exceptions recognized and handled by ABP Framework.

In the simplest case, you can directly use the BusinessException class to throw a business exception. See the following example from the *EventHub* project:

```
public class EventRegistrationManager : DomainService
{
    public async Task RegisterAsync(
        Event @event,
        AppUser user)
    {
        if (Clock.Now > @event.EndTime)
        {
            throw new BusinessException(EventHubErrorCodes
                .CantRegisterOrUnregisterForAPastEvent);
        }
        ...
    }
}
```

`EventRegistrationManager` is a domain service that is used to perform business rules for event registrations. The `RegisterAsync` method checks the event time and prevents registering to events in the past by throwing a business exception in that case.

The constructor of `BusinessException` takes a few parameters, and all are optional. These are listed here:

- `code`: A string value that is used as a custom error code for the exception. Client applications can check it while handling the exception and track the error type easily. You typically use different error codes for different exceptions. The error code can also be used to localize the exception, as we will see in the *Localizing a business exception* section.

- `message`: A string exception message, if needed.

- `details`: A detailed explanation message string, if needed.

- `innerException`: An inner exception, if available. You can pass here if you have cached an exception and throw a business exception based on that exception.

- `logLevel`: The logging level for this exception. It is an enum of the `LogLevel` type, and the default value is `LogLevel.Warning`.

You generally only pass `code`, which is easier to find in logs. It is also used for localizing the error message returned to the client.

Localizing a business exception

If you use `UserFriendlyException`, you have to localize the message yourself since the exception message is shown directly to the end user. If you throw `BusinessException`, ABP doesn't show the exception message to the end user unless you explicitly localize it. It uses error code namespaces for that purpose.

Assume that you've used `EventHub:CantRegisterOrUnregisterForAPastEvent` as the error code. `EventHub`, here, becomes the error code namespace through the usage of the colon. We must map the error code namespace to a localization resource so that ABP can know which localization resource to use for these error messages. The code is illustrated in the following snippet:

```
Configure<AbpExceptionLocalizationOptions>(options =>
{
    options.MapCodeNamespace(
        "EventHub", typeof(EventHubResource));
});
```

In this code snippet, we map the `EventHub` error code namespace to the `EventHubResource` localization resource. Now, you can define the error code as a key in your localization file, including the namespace, as follows:

```
{
    "culture": "en",
    "texts": {
        "EventHub:CantRegisterOrUnregisterForAPastEvent":
            "You can not register to or unregister from an
            event in the past, sorry!"
    }
}
```

After that configuration, ABP shows the localized message to the user whenever you throw a `BusinessException` exception with that error code.

In some cases, you may want to include some additional data in the error message. See the following code snippet:

```
throw new BusinessException(
    EventHubErrorCodes.OrganizationNameAlreadyExists
).WithData("Name", name);
```

Here, we include the organization name in the error message, using the `WithData` extension method. Then, we can define the localization string, as shown in the following code snippet:

```
"EventHub:OrganizationNameAlreadyExists": "The organization
{Name} already exists. Please use another name."
```

In this example, {Name} is a placeholder for the organization name. ABP automatically replaces it with the given name.

We will cover the localization system in *Chapter 8, Using the Features and Services of ABP*.

We've seen how to throw a `BusinessException` exception. What if you want to create specialized exception classes?

Custom business exception classes

It is also possible to create custom exception classes instead of directly throwing a BusinessException exception. In this case, you can create a new class inheriting from the BusinessException class, as shown in the following code block:

```
public class OrganizationNameAlreadyExistsException
    : BusinessException
{
    public string Name { get; private set; }

    public OrganizationNameAlreadyExistsException(
        string name) : base(EventHubErrorCodes
        .OrganizationNameAlreadyExists)
    {
        Name = name;
        WithData("Name", name);
    }
}
```

In this example, OrganizationNameAlreadyExistsException is a custom business exception class. It takes the organization's name in its constructor. It sets the "Name" data so that ABP can use the organization name in the localization process. Throwing this exception is pretty straightforward, as we can see here:

```
throw new OrganizationNameAlreadyExistsException(name);
```

This usage is simpler than throwing a BusinessException exception with custom data, which the developer can forget to set. It also reduces duplication when you throw the same exception in multiple places in your code base.

Controlling exception logging

As mentioned at the beginning of the *Exception handling* section, ABP automatically logs all exceptions. Business exceptions, authorization, and validation exceptions are logged with the Warning level, while other errors are logged with the Error level by default.

You can implement the `IHasLogLevel` interface to set a different log level for an exception class. See the following example:

```
public class MyException : Exception, IHasLogLevel
{
    public LogLevel LogLevel { get; set; } =
        LogLevel.Warning;


    //...
}
```

The `MyException` class implements the `IHasLogLevel` interface with the `Warning` level. ABP will write warning logs if you throw exceptions of the `MyException` type.

It is also possible to write additional logs for an exception. You can implement the `IExceptionWithSelfLogging` interface to write additional logs, as shown in the following example:

```
public class MyException
    : Exception, IExceptionWithSelfLogging
{
    public void Log(ILogger logger)
    {
        //...log additional info
    }
}
```

In this example, the `MyException` class implements the `IExceptionWithSelfLogging` interface, which defines a `Log` method. ABP passes the logger here to allow you to write additional logs if you need them.

Controlling the HTTP status code

ABP does its best to return a proper HTTP status code for known exception types, as follows:

- Returns `401` (unauthorized) if the user has not logged in, for `AbpAuthorizationException`

- Returns `403` (forbidden) if the user has logged in, for `AbpAuthorizationException`

- Returns `400` (bad request) for `AbpValidationException`

- Returns `404` (not found) for `EntityNotFoundException`

- Returns `403` (forbidden) for business and user-friendly exceptions

- Returns `501` (not implemented) for `NotImplementedException`

- Returns `500` (internal server error) for other exceptions (those are assumed to be infrastructure errors)

If you want to return another HTTP status code for your custom exceptions, you can map your error code to an HTTP status code, as shown in the following configuration:

```
services.Configure<AbpExceptionHttpStatusCodeOptions>(
    options =>
{
    options.Map(
        EventHubErrorCodes.OrganizationNameAlreadyExists,
        HttpStatusCode.Conflict);
});
```

It is suggested to make that configuration in the web or HTTP API layer of your solution.

Summary

In this chapter, we've explored three fundamental cross-cutting concerns that we should implement in every serious business application.

Authorization is a key concern for system security. You should carefully control the authorization rules in every operation of your application. ABP simplifies the use of ASP.NET Core's authorization infrastructure and adds a flexible permission system that is a very common pattern for enterprise applications.

Validation, on the other hand, supports system security and improves the UX by gracefully preventing badly formatted or malicious requests. ABP enhances the standard ASP.NET Core validation by allowing you to implement the validation in any service of your application and integrating it into the `FluentValidation` library for advanced usage.

Finally, ABP's exception-handling system works seamlessly and automates exception handling on the server side and client side. It also allows you to decouple localizing error messages and map them to HTTP status codes from your code that throws an exception.

The next chapter will continue to explore ABP Framework services by introducing some fancy ABP features such as automatic audit logging and data filtering.

8
Using the Features and Services of ABP

ABP Framework is a full-stack application development framework, so it has many building blocks for every aspect of an enterprise solution. In the last three chapters, we have explored the fundamental services, data access infrastructure, and cross-cutting concern solutions provided by ABP Framework.

In this final chapter of *Part 2, Fundamentals of ABP Framework*, we will continue with some ABP features frequently used in business applications, in the following order:

- Obtaining the current user
- Using the data filtering system
- Controlling the audit logging system
- Caching data
- Localizing the **user interface (UI)**

Technical requirements

If you want to follow and try the examples, you need to install an **integrated development environment (IDE)**/editor (such as Visual Studio) to build ASP.NET Core projects.

You can download the code examples from the following GitHub repository: `https://github.com/PacktPublishing/Mastering-ABP-Framework`.

Obtaining the current user

If your application requires user authentication for some functionalities, you generally need to get information about the current user. ABP provides the `ICurrentUser` service to obtain detailed information for the currently logged-in user. For web applications, the implementation of `ICurrentUser` is completely integrated with ASP.NET Core's authentication system, so you can easily get claims of the current user.

See the following code block for simple usage of the `ICurrentUser` service:

```
using System;
using Volo.Abp.DependencyInjection;
using Volo.Abp.Users;
namespace DemoApp
{
    public class MyService : ITransientDependency
    {
        private readonly ICurrentUser _currentUser;

        public MyService(ICurrentUser currentUser)
        {
            _currentUser = currentUser;
        }

        public void Demo()
        {
            Guid? userId = _currentUser.Id;
            string userName = _currentUser.UserName;
            string email = _currentUser.Email;
        }
    }
}
```

In this example, the `MyService` constructor injects the `ICurrentUser` service, then gets the unique `Id`, `Username`, and `Email` values of the current user.

Here are the properties of the `ICurrentUser` interface:

- `IsAuthenticated` (`bool`): Returns `true` if the current user has logged in (authenticated).

- `Id` (`Guid?`): The **unique identifier** (**UID**) of the current user. Returns `null` if the current user has not logged in.

- `UserName` (`string`): Username of the current user. Returns `null` if the current user has not logged in.

- `TenantId` (`Guid?`): Tenant ID of the current user. It is usable for multi-tenant applications. Returns `null` if the current user is not related to a tenant.

- `Email` (`string`): Email address of the current user. Returns `null` if the current user has not logged in or has not set an email address.

- `EmailVerified` (`bool`): Returns `true` if the current user's email address has been verified.

- `PhoneNumber` (`string`): Phone number of the current user. Returns `null` if the current user has not logged in or has not set a phone number.

- `PhoneNumberVerified` (`bool`): Returns `true` if the current user's phone number has been verified.

- `Roles` (`string[]`): All roles of the current user as a string array.

> **Injecting the ICurrentUser Service**
>
> `ICurrentUser` is a widely used service. Thus, some base ABP classes (such as `ApplicationService` and `AbpController`) provide it pre-injected. In these classes, you can directly use the `CurrentUser` property instead of manually injecting this service.

ABP can work with any authentication provider since it works with the current claims that are provided by ASP.NET Core. **Claims** are key-value pairs issued on user login and stored in the authentication ticket. If you are using cookie-based authentication, they are stored in a cookie and sent to the server in every request. If you are using token-based authentication, they are sent by the client in every request, typically in the **HyperText Transfer Protocol** (**HTTP**) header.

The `ICurrentUser` service gets all the information from the current claims. If you want to query the current claims directly, you can use the `FindClaim`, `FindClaims`, and `GetAllClaims` methods. These methods are especially useful if you create your own custom claims.

Defining custom claims

ABP provides an easy way to add your custom claims to the authentication ticket so that you can safely get these custom values on the next requests of the same user. You can implement the `IAbpClaimsPrincipalContributor` interface to add custom claims to the authentication ticket.

In the following example, we are adding social security number information—a custom claim—to the authentication ticket:

```
public class SocialSecurityNumberClaimsPrincipalContributor
    : IAbpClaimsPrincipalContributor, ITransientDependency
{
    public async Task ContributeAsync(
        AbpClaimsPrincipalContributorContext context)
    {
        ClaimsIdentity identity = context.ClaimsPrincipal
            .Identities.FirstOrDefault();
        var userId = identity?.FindUserId();
        if (userId.HasValue)
        {
            var userService = context.ServiceProvider
                .GetRequiredService<IUserService>();
            var socialSecurityNumber = await userService
                .GetSocialSecurityNumberAsync(userId.Value);
            if (socialSecurityNumber != null)
            {
                identity.AddClaim(new Claim
                    ("SocialSecurityNumber",
                        socialSecurityNumber));
            }
        }
    }
}
```

In this example, we are first getting the `ClaimsIdentity` and finding the current user's ID. Then, we are getting the social security number from `IUserService`, which is a custom service that you should develop yourself. You can get any service from the `ServiceProvider` to query the data that you need. Finally, we are adding a new `Claim` to the `identity`. `SocialSecurityNumberClaimsPrincipalContributor` is then used whenever a user logs in to the application.

You can use custom claims to authorize the current user for specific business requirements, filter data, or just show on the UI. Notice that authentication ticket claims cannot be changed unless you invalidate the authentication ticket and force the user to re-authenticate, so do not store frequently changed data in the claims. You can use the caching system (which will be introduced in the *Caching data* section) if your purpose is to store user data where it can be quickly accessed later.

`ICurrentUser` is a core service that you frequently use in your application code. The next section introduces the data filtering system that seamlessly works most of the time.

Using the data filtering system

Filtering data in a query is very common in database operations. If you are using **Structured Query Language** (**SQL**), you can use the `WHERE` clause. If you are using **Language Integrated Query** (**LINQ**), you use the `Where` extension method in C#. While most of these filtering conditions vary in your queries, some expressions are applied to all queries you run if you implement patterns such as soft-delete and multi-tenancy.

ABP automates the data filtering process to help you avoid repeating the same filtering logic everywhere in your application code.

In this section, we will first see the pre-built data filters of ABP Framework, then learn how to disable the filters when we need to. Finally, we will see how to implement our custom data filters.

We typically use simple interfaces to enable filtering for entities. ABP defines two pre-defined data filters to implement soft-delete and multi-tenancy patterns.

The soft-delete data filter

If you use the soft-delete pattern for an entity, you never delete the entity in the database physically. Instead, you mark it as *deleted*.

ABP defines the ISoftDelete interface to standardize the property to mark an entity as soft-delete. You can implement that interface for an entity, as shown in the following code block:

```
public class Order : AggregateRoot<Guid>, ISoftDelete
{
    public bool IsDeleted { get; set; }
    //...other properties
}
```

The Order entity, in this example, has an IsDeleted property that is defined by the ISoftDelete interface. Once you implement that interface, ABP automates the following tasks for you:

- When you delete an order, ABP identifies that the Order entity implements the soft-delete pattern, prevents the deletion, and sets IsDeleted to true. So, the order is not physically deleted in the database.

- When you query orders, ABP automatically filters deleted entities (by adding an IsDeleted == false condition to the query) to avoid accidentally retrieving deleted orders from the database.

Data filtering is related to queries, so, the first task is not directly related to data filtering but is a supporting logic implemented by ABP Framework.

> **Data Filtering Limitations**
>
> The data filtering automation only works when you use repositories or DbContext (for **Entity Framework Core (EF Core)**). Otherwise, for example, if you are using a hand-written SQL DELETE or SELECT command, you should handle this yourself because ABP cannot intercept your operation in such cases.

The soft-delete filter is one of the built-in ABP data filters. Another built-in filter is for multi-tenancy.

The multi-tenancy data filter

Multi-tenancy is a widely used pattern to share resources between tenants in **software-as-a-service (SaaS)** solutions. It is essential to isolate the data between different tenants in a multi-tenant application. One tenant cannot read or write to another tenant's data, even if they are located in the same physical database.

ABP has a complete multi-tenant system, which will be explained in detail in *Chapter 16, Implementing Multi-Tenancy*. However, it would be good to mention the multi-tenancy filter here since it is related to the data filtering system.

ABP defines the `IMultiTenant` interface to enable the multi-tenancy data filter for an entity. We can implement that interface for an entity, as shown in the following code block:

```
public class Order : AggregateRoot<Guid>, IMultiTenant
{
    public Guid? TenantId { get; set; }
    //...other properties
}
```

The `IMultiTenant` interface defines the `TenantId` property, as shown in this example. ABP uses `Guid` values for tenant IDs.

Once we implement the `IMultiTenant` interface, ABP automatically filters all queries for the `Order` entity using the ID of the current tenant. The current tenant's ID is obtained from the `ICurrentTenant` service, which will be explained in *Chapter 16, Implementing Multi-Tenancy*.

Working with Multiple Data Filters

Multiple data filters can be enabled for the same entity. For example, the `Order` entity defined in this section could implement both the `ISoftDelete` and `IMultiTenant` interfaces.

As you see, implementing a data filter for an entity is pretty easy—just implement the interface related to the data filter. All data filters are enabled by default unless you explicitly disable them.

Disabling a data filter

Disabling an automatic filter can be necessary in some cases—for example, you may want to disable the soft-delete filter to read deleted entities from the database, or maybe you want to allow the user to recover deleted entities. You may want to disable the multi-tenancy filter to query data from all the tenants in a multi-tenant system. For whatever reason, ABP provides an easy and safe way to disable a data filter.

The following example shows how to get all the orders from the database, including deleted ones, by disabling the `ISoftDelete` data filter using the `IDataFilter` service:

```
public class OrderService : ITransientDependency
{
    private readonly IRepository<Order, Guid>
    _orderRepository;
    private readonly IdataFilter _dataFilter;

    public OrderService(
        Irepository<Order, Guid> orderRepository,
        IdataFilter dataFilter)
    {
        _orderRepository = orderRepository;
        _dataFilter = dataFilter;
    }

    public async Task<List<Order>> GetAllOrders()
    {
        using (_dataFilter.Disable<IsoftDelete>())
        {
            return await _orderRepository.GetListAsync();
        }
    }
}
```

`OrderService`, in this example, injects the `Order` repository and the `IdataFilter` service. It then uses the `_dataFilter.Disable<IsoftDelete>()` expression to disable the soft-delete filter. In the `using` statement, the filter is disabled, and we can query deleted orders too.

> **Always Use a using Statement**
>
> The `Disable` method returns a disposable object so that we can use it in a `using` statement. Once the `using` block ends, the filter automatically turns back to the previous state, which means that if it was enabled before that `using` block, it returns to the enabled state. If it was already disabled before the `using` statement, the `Disable` method does not affect it, and it remains disabled after the `using` statement. This system allows us to safely disable a filter without affecting any logic that calls the `GetAllOrders` method. It is always recommended to disable a filter in a `using` statement.

`IdataFilter` service provides two more methods:

- `Enable<Tfilter>`: Enables a data filter. You can use this to temporarily enable a data filter in a scope in which the filter was disabled. It has no effect if the filter is already enabled. It is always recommended to enable a filter in a `using` statement, just as with the `Disable` method.

- `IsEnabled<Tfilter>`: Returns `true` if the given filter is currently enabled. You generally do not need this method since `Enable` and `Disable` work as expected in both cases.

We've learned how to use the `Disable` and `Enable` pre-built data filters. The next section shows how to create custom data filters.

Defining custom data filters

Just as with pre-built data filters, you may want to define your own filters. A data filter is represented by an interface, so the first step is to define an interface for your filter.

Assume that you want to archive your entities and automatically filter the archived data to not retrieve them into the application by default. For this example, we can define such a simple interface (you can define this in your domain layer), as follows:

```
public interface Iarchivable
{
    bool IsArchived { get; }
}
```

The IsArchived property will be used to filter the entities. Entities with IsArchived is true will be eliminated by default. Once we define such an interface, we can implement it for the entities that can be archived. See the following example:

```
public class Order : AggregateRoot<Guid>, Iarchivable
{
    public bool IsArchived { get; set; }
    //...other properties
}
```

The Order entity, in this example, implements the Iarchivable interface, which makes it possible to apply the data filter on that entity.

Note that the Iarchivable interface doesn't define a setter for IsArchived, but the Order entity defines it. That is my design decision; we don't need to set IsArchived over the interface, but we need to set it on the entity.

Since data filtering is done at the database provider level, custom filter implementation also depends on the database provider. This section will show how to implement the Iarchivable filter for the EF Core provider. If you are looking for MongoDB, please refer to ABP's documentation: https://docs.abp.io/en/abp/latest/Data-Filtering.

ABP uses EF Core's **Global Query Filters** system for data filtering in EF Core. You can implement filtering logic for the data filter in your DbContext class.

The first step is to define a property in your DbContext class that will be used in the filter expression, as follows:

```
protected bool IsArchiveFilterEnabled => DataFilter?.
IsEnabled<Iarchivable>() ?? false;
```

This property directly uses the IdataFilter service to get the filter state. The DataFilter property comes from the base AbpDbContext class, and it can be null if the DbContext instance was not resolved from the **dependency injection** (**DI**) system. That's why I've used the null check.

The next step is to override the ShouldFilterEntity method to decide if a given entity type should be filtered or not:

```
protected override bool ShouldFilterEntity<Tentity>(
    ImutableEntityType entityType)
{
```

```
    If (typeof(IArchivable)
        .IsAssignableFrom(typeof(TEntity)))
    {
        return true;
    }

    return base.ShouldFilterEntity<TEntity>(entityType);
}
```

ABP Framework calls this method for each entity type in this DbContext class (it is called only once—the first time the DbContext class is used after an application start). If this method returns true, it enables the EF Core global filters for that entity. Here, I just checked if the given entity implemented the IArchivable interface and returned true in that case. Otherwise, call the base method so that it checks for other data filters.

ShouldFilterEntity only decides to enable filtering or not. The actual filtering logic should be implemented by overriding the CreateFilterExpression method:

```
protected override Expression<Func<TEntity, bool>>
CreateFilterExpression<TEntity>()
{
    var expression =
        base.CreateFilterExpression<Tentity>();
    if (typeof(Iarchivable)
        .IsAssignableFrom(typeof(TEntity)))
    {
        Expression<Func<TEntity, bool>> archiveFilter =
            e => !IsArchiveFilterEnabled ||
                !EF.Property<bool>(e, "IsArchived");
        expression = expression == null
            ? archiveFilter
            : CombineExpressions(expression,
                archiveFilter);
    }
    return expression;
}
```

The implementation seems a bit complicated because it creates and combines expressions. The important part is how the `archiveFilter` expression was defined. `!IsArchiveFilterEnabled` checks if the filter is disabled. If the filter is disabled, then the other condition is not evaluated, and all the entities are retrieved without filtering. `!EF.Property<bool>(e, "IsArchived")` checks if the `IsArchived` value is `false` for that entity, so it eliminates entities with `IsArchived` as `true`.

As you've seen from the preceding code block, I haven't used the `Order` entity in the filter implementation. That means the implementation is generic and can work with any entity type—all you need is to implement the `IArchivable` interface for the entity that you want to apply the filter for.

In summary, ABP allows us to create and control global query filters easily. It also uses that system to implement two popular patterns—soft-delete and multi-tenancy. The next section introduces the audit logging system, ABP's other feature that is very common in enterprise software solutions.

Controlling the audit logging system

ABP's audit logging system tracks all requests and entity changes and writes them into a database. Then, you can get a report of what was done in your application, when it was made, and who did it.

The audit log system is installed and properly configured when you create a new solution from the startup templates. Most of the time, you use it without any configuration. However, ABP allows you to control, customize, and extend the audit logging system. But first, let's understand what an audit log object is.

Audit log object

An audit log object is a group of actions and related entity changes performed together in a limited scope, typically in an HTTP request for a web application. We will talk more about audit log scopes in the next section.

The diagram in *Figure 8.1* represents an audit log object:

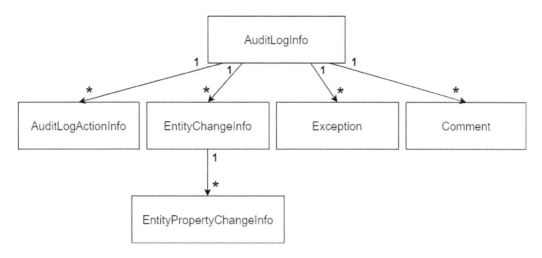

Figure 8.1 – Audit log object

Let's explain that diagram by beginning from the root object, as follows:

- `AuditLogInfo`: In every scope (typically, a web request), there is one `AuditLogInfo` object that contains information about the current user, current tenant, HTTP request, client and browser details, and execution time and duration of the operation.

- `AuditLogActionInfo`: In every audit log, there may be zero or more actions. An action is typically a controller action call, a page handler call, or an application service method call. It includes the class name, method name, and method arguments in that call.

- `EntityChangeInfo`: An audit log object may contain zero or more changes on the entities in the database. Each entity change contains the change type (created, updated, or deleted), entity type (full class name), and ID of the changed entity.

- `EntityPropertyChangeInfo`: For every entity change, it logs the changes on the properties (fields in the database). This object contains the name, type, old value, and the new value of the affected property.

- `Exception`: A list of exceptions occurred during this audit log scope.

- `Comment`: Additional comments/logs related to this audit log.

The audit log object is saved into multiple tables in a relational database: `AbpAuditLogs`, `AbpAuditLogActions`, `AbpEntityChanges`, and `AbpEntityPropertyChanges`. I've written the fundamental properties of the audit log object in the previous list. You can check these database tables or investigate the `AuditLogInfo` object to see all the details.

> **MongoDB Limitation**
>
> Entity changes are not logged for MongoDB since ABP uses EF Core's change-tracking system to get the entity change information, and the MongoDB driver has no such change-tracking system.

As mentioned at the beginning of this section, an audit log object is created per audit log scope.

Audit log scope

The audit log scope uses the **Ambient Context Pattern**. When you create a new audit log scope, all actions and changes made in this scope are saved as a single audit log object.

There are a few ways to establish an audit log scope.

Audit log middleware

The first and the most common way to create an audit log scope is to use the audit log middleware in the ASP.NET Core pipeline configuration:

```
app.UseAuditing();
```

This is typically placed before the `app.UseEndpoints()` or `app.UseConfiguredEndpoints()` endpoint configuration. When you use this middleware, every HTTP request writes a separate audit log record, which is the wanted behavior most of the time and is already configured in the startup templates by default.

Audit log interceptor

If you don't use the audit log middleware or if your application is not a request/reply-style ASP.NET Core application (for example, a desktop or Blazor Server application), then ABP creates a new audit log scope per application service method.

Manually creating audit scopes

You typically won't need to do this, but if you want to create an audit scope manually, you can use the `IAuditingManager` service, as shown in the following code block:

```
public class MyServiceWithAuditing : ITransientDependency
{
    //...inject IAuditingManager _auditingManager;
    public async Task DoItAsync()
```

```
    {
        using (var auditingScope =
            _auditingManager.BeginScope())
        {
            try
            {
                //TODO: call other services...
            }
            catch (Exception ex)
            { _auditingManager.Current.Log.Exceptions.Add(ex);
                throw;
            }
            finally
            {
                await auditingScope.SaveAsync();
            }
        }
    }
}
```

Once you inject the `IAuditingManager` service, you can use the `BeginScope` method to create a new scope. Then, create a `try-catch` block to save the audit log, including exception cases. In the `try` section, you can just perform your logic, call any other service, and so on. All these operations and the changes in these operations are saved as a single audit log object in the `finally` block.

Inside an audit log scope (regardless of whether it is created by ABP or manually by you), `_auditingManager.Current.Log` can be used to get the current audit log object to investigate or manipulate it (for example, add comment lines or additional information). If you are not in an audit log scope, then `_auditingManager.Current` returns `null`, so check for `null` if you are unsure as to whether there is a surrounding audit log scope.

I've introduced the audit log object and audit log scopes, which work seamlessly by default. Now, let's see options to understand the defaults and the global configuration possibilities for the audit log system.

Auditing options

The `AbpAuditingOptions` class is used to configure default options for the auditing system. It can be configured using the standard `options` pattern, as shown in the following example:

```
Configure<AbpAuditingOptions>(options =>
{
    options.IsEnabled = false;
});
```

You can configure `options` inside the `ConfigureServices` method of your module. See the following list for the main options for the auditing system:

- `IsEnabled` (bool; default: `true`): The main point to completely disable the auditing system.

- `IsEnabledForGetRequests` (bool; default: `false`): ABP does not save audit logs for HTTP GET requests by default because GET requests are not supposed to change the database. However, you can set this to `true`, which enables audit logging for GET requests too.

- `IsEnabledForAnonymousUsers` (bool; default: `true`): Set this to `false` if you want to write audit logs only for authenticated users. If you save audit logs for anonymous users, you will see `null` for `UserId` values for these users.

- `AlwaysLogOnException` (bool; default: `true`): If an exception occurs in your application code, ABP saves the audit log by default, without considering the `IsEnabledForGetRequests` and `IsEnabledForAnonymousUsers` options. Set this to `false` to disable that behavior.

- `hideErrors` (bool; default: `true`): ABP ignores exceptions while saving audit log objects to the database. Set this to `false` to throw exceptions instead of hiding them.

- `ApplicationName` (string; default: `null`): If multiple applications are using the same database to save the audit logs, you can set this option in each application so that you can filter the logs based on the application name.

- `IgnoredTypes` (List<Type>): You can ignore some specific types in the audit log system, including entity types.

In addition to these simple global options, you can enable/disable change tracking for entities.

Enabling entity histories

The audit log object contains entity changes with property details. However, it is disabled for all entities by default because it may write too many logs into the database, which may rapidly increase the database size. It is suggested to enable it in a controlled way for the entities you want to track.

There are two ways to enable entity histories for entities, as outlined here:

- The [Auditing] attribute is used to enable it for a single entity. It will be explained in the next section.

- The EntityHistorySelectors option is used to enable it for multiple entities.

In the following example, I've enabled the EntityHistorySelectors option for all entities:

```
Configure<AbpAuditingOptions>(options =>
{
    options.EntityHistorySelectors.AddAllEntities();
});
```

The AddAllEntities method is a shortcut. EntityHistorySelectors is a list of named selectors, and you can add a lambda expression to select the entities you want. The following code is equivalent to the preceding configuration code:

```
Configure<AbpAuditingOptions>(options =>
{
    options.EntityHistorySelectors.Add(
        new NamedTypeSelector("MySelectorName", type =>
            true)
    );
});
```

The first argument of NamedTypeSelector is the selector name—MySelectorName, for this example. Selector names are arbitrary, and they can be used later to find or remove a selector from the selector list. You typically don't use it; just give it a unique name. The second argument of NamedTypeSelector takes an expression. It gives you an entity type and waits for true or false. Returns true, if you want to enable entity histories for a given entity type. So, you can pass an expression such as type => type.Namespace.StartsWith("MyRootNamespace") to select all entities with a namespace. You can add as many selectors as you need. All selectors are tested. If one of them returns true, the entity is selected for logging property changes.

Besides these global options and selectors, there are ways to enable/disable audit logging per class, method, and property level.

Disabling and enabling audit logging in detail

When you use the audit log system, you typically want to log every access. However, in some cases, you may want to disable audit logging for some specific actions or entities. Here are some potential reasons for that: the action parameters may be dangerous to write into the logs (for example, it may contain the user's password), the action call or entity change might be out of the user's control, so it isn't worth recording for audit purposes, or the operation can be a bulk operation that writes too many audit logs and decreases performance.

ABP defines the [DisableAuditing] and [Audited] attributes to control logged objects declaratively. There are two targets that you can control for audit logging: service calls and entity histories.

Controlling audit logging for service calls

Application service methods, Razor Page handlers, and **model-view-controller** (**MVC**) controller actions are included in the audit log by default. To disable audit logging for them, you can use the [DisableAuditing] attribute at the class or method level.

The following example uses the [DisableAuditing] attribute on an application service class:

```
[DisableAuditing]
public class OrderAppService : ApplicationService,
IOrderAppService
{
    public async Task CreateAsync(CreateOrderDto input)
    {
    }
    public async Task DeleteAsync(Guid id)
    {
    }
}
```

With this usage, ABP won't include the execution of these methods into the audit log object. If you just want to disable one of the methods, you can use it at the method level:

```
public class OrderAppService : ApplicationService,
IOrderAppService
{
    [DisableAuditing]
    public async Task CreateAsync(CreateOrderDto input)
    {
    }
    public async Task DeleteAsync(Guid id)
    {
    }
}
```

In this case, the CreateAsync method call is not included in the audit log, while the DeleteAsync method call is written into the audit log object. The same behavior could be accomplished using the following code:

```
[DisableAuditing]
public class OrderAppService : ApplicationService,
IOrderAppService
{
    public async Task CreateAsync(CreateOrderDto input)
    {
    }
    [Audited]
    public async Task DeleteAsync(Guid id)
    {
    }
}
```

I disabled audit logging for all methods except the DeleteAsync method because the DeleteAsync method declares the [Audited] attribute.

The [Audited] attribute can be used on any class (used with the DI system) to enable audit logging on that class, even if the class is not audit-logged by default. Moreover, you can use it in any method of any class to just enable it for that particular method call. If you use the [Audited] attribute on a class, you can then disable a specific method using the [DisableAuditing] attribute.

When ABP includes a method call information in the audit log object, it also includes all the parameters of the executed method. That is super-useful to understand which changes were made in your system; however, you may want to exclude some properties of the input in some cases. Consider a scenario that you get credit card information from the user. You probably don't want to include this in the audit logs. You can use the [DisableAuditing] attribute on any property of an input object in such a case. See the following example, which excludes a property of a Dto input from the audit log:

```
public class CreateOrderDto
{
    public Guid CustomerId { get; set; }
    public string DeliveryAddress { get; set; }

    [DisableAuditing]
    public string CreditCardNumber { get; set; }
}
```

For this example, ABP won't write the CreditCardNumber value into the audit log.

Disabling audit logging for method calls won't affect the entity histories. If an entity is changed and it was selected for audit logging, changes are still logged. The next section explains how to control the audit logging system for entity histories.

Controlling audit logging for entity histories

In the *Enabling entity histories* section, we saw how to enable entity histories for one or more entities by defining selectors. However, if you want to enable the entity histories for a single entity, there is an alternative and simpler way: just add the [Audited] attribute above your entity class:

```
[Audited]
public class Order : AggregateRoot<Guid>
{
}
```

In this example, I added the [Audited] attribute to an Order entity to configure the audit logging system to enable entity histories for this entity.

Suppose you've used selectors to enable entity histories for many or all entities but want to disable them for a specific entity. In that case, you can use the [DisableAuditing] attribute for that entity class.

The [DisableAuditing] attribute can also be used on a property of an entity to exclude this property from the audit logs, as shown in the following example:

```
[Audited]
public class Order : AggregateRoot<Guid>
{
    public Guid CustomerId { get; set; }

    [DisableAuditing]
    public string CreditCardNumber { get; set; }
}
```

For that example, ABP won't write the CreditCardNumber value into the audit log.

Storing audit logs

The core of the ABP Framework has been designed to not assume any data store by introducing abstractions wherever it needs to touch a data source. The audit logging system is not an exception. It defines the IAuditingStore interface to abstract where the audit log objects are saved. That interface only has a single method:

```
Task SaveAsync(AuditLogInfo auditInfo);
```

You can implement this interface to save audit logs where you want. If you use ABP's startup templates to create a new solution, it is configured to save audit logs to the application's main database, so you normally don't have to implement the IAuditingStore interface manually.

We've seen different ways to control and customize the audit log system. Audit logging is an essential system for enterprise systems to track and log changes in your system. The next section introduces the caching system, another essential feature of a web application.

Caching data

Caching is one of the most fundamental systems to improve your application's performance and scalability. ABP extends ASP.NET Core's **distributed caching** system and makes it compatible with other features of ABP Framework, such as multi-tenancy.

Distributed caching is essential if you run multiple instances of your application or have a distributed system, such as a microservice solution. It provides consistency between different applications and allows the sharing of cached values. A distributed cache is typically an external, standalone application, such as Redis and Memcached.

It is suggested to use the distributed caching system even if your application has a single running instance. Don't worry about the performance since the default implementation of distributed cache works in memory. That means it is not distributed unless you explicitly configure a real distributed cache provider, such as Redis.

Distributed Caching in ASP.NET Core

This section focuses on ABP's caching features and doesn't cover all ASP. NET Core's distributed caching system features. You can refer to Microsoft's documentation to learn more about distributed caching in ASP.NET Core: `https://docs.microsoft.com/en-us/aspnet/core/ performance/caching/distributed`.

In this section, I will show you how to use the `IDistributedCache<T>` interface, configure options, and deal with error handling and batch operations. We will also learn about using Redis as the distributed cache provider. Finally, I will talk about invalidating cached values.

Let's start from the basics—the `IDistributedCache<T>` interface.

Using the IDistributedCache<T> interface

ASP.NET Core defines an `IDistributedCache` interface, but it is not type-safe. It sets and gets `byte` arrays rather than objects. ABP's `IDistributedCache<T>` interface, on the other hand, is designed as generic with type-safe method parameters (`T` stands for the type of items stored in the cache). It internally uses the standard `IDistributedCache` interface to be 100% compatible with ASP.NET Core's caching system. ABP's `IDistributedCache<T>` interface has two main advantages, as follows:

- Automatically serializes/deserializes the objects to **JavaScript Object Notation (JSON)** values, then to `byte` arrays. So, you don't deal with serialization and deserialization.

- It automatically adds the cache name prefix to the cache keys to allow the use of the same key for different kinds of cache objects.

The first step in using the `IDistributedCache<T>` interface is to define a class to represent items in the cache. I've defined the following class to store a user's information in the cache:

```
public class UserCacheItem
{
    public Guid Id { get; set; }
```

```
    public string UserName { get; set; }
    public string EmailAddress { get; set; }
}
```

That is a plain C# class. The only restriction is that it should be serializable because it is serialized to JSON while saving to the cache and deserialized while reading from the cache (for example, do not add references to other objects that should not or cannot be stored in the cache; keep it simple).

Once we've defined the cache item class, we can inject the `IDistributedCache<T>` interface, as shown in the following code block:

```
public class MyUserService : ITransientDependency
{
    private readonly IDistributedCache<UserCacheItem>
        _userCache;

    public MyUserService(IDistributedCache<UserCacheItem>
        userCache)
    {
        _userCache = userCache;
    }
}
```

I've injected the `IDistributedCache<UserCacheItem>` service to work with the distributed cache for `UserCacheItem` objects. The following code block shows how we can use it to get cached user information and fall back to the database query if the given user was not found in the cache:

```
public async Task<UserCacheItem> GetUserInfoAsync(Guid userId)
{
    return await _userCache.GetOrAddAsync(
        userId.ToString(),
        async () => await GetUserFromDatabaseAsync(userId),
        () => new DistributedCacheEntryOptions
        {
            AbsoluteExpiration =
                DateTimeOffset.Now.AddHours(1)
        }
```

```
    );
}
```

I've passed three parameters to the `GetOrAddAsync` method:

- The first parameter is the cache key, which should be a string value, so I converted the `Guid userId` value to a string value.

- The second parameter is a factory method that is executed if the given key is not found in the cache. I passed the `GetUserFromDatabaseAsync` method here. In that method, you should build the cache item from its data source.

- The final parameter is a factory method that returns a `DistributedCacheEntryOptions` object. This is optional and configures the expiration time for the cached item. The factory method is only called if the `GetOrAddAsync` method adds the entry.

Cache keys are `string` data types by default. However, ABP defines another interface, `IDistributedCache<TCacheItem, TCacheKey>`, allowing you to specify the cache key so that you don't need to convert your keys to `string` data types manually. We could inject the `IDistributedCache<UserCacheItem, Guid>` service and remove the `ToString()` usage in the first parameter for this example.

`DistributedCacheEntryOptions` has the following options to control the lifetime of the cached item:

- `AbsoluteExpiration`: You can set an absolute time, as we've done in this example. The item is automatically deleted from the cache at that time.

- `AbsoluteExpirationRelativeToNow`: An alternative way to set the absolute expiration time. We could rewrite the option in this example so that it reads `AbsoluteExpirationRelativeToNow = TimeSpan.FromHours(1)`. The result will be the same.

- `SlidingExpiration`: Sets how long the cache item can be inactive (not accessed) before it is removed. This means that if you continue to access the cached item, the expiration time is automatically extended.

If you don't pass the expiration time parameter, the default value is used. You can configure the default value and some other global options using the `AbpDistributedCacheOptions` class explained in the next section. Before that, let's see the other methods of the `IDistributedCache<UserCacheItem>` service, as follows:

- `GetAsync` is used to read data from the cache with a cache key.
- `SetAsync` is used to save an item to the cache. It overwrites the existing value if available.
- `RefreshAsync` is used to reset the sliding expiration time for the given key.
- `RemoveAsync` is used to delete an item from the cache.

> **About Synchronous Cache Methods**
>
> All the methods also have synchronous versions, such as the GET method for the `GetAsync` method. However, it is suggested to use the asynchronous version wherever possible.

These methods are the standard methods of ASP.NET Core. ABP adds methods to work with multiple items for each method, such as `GetManyAsync` for `GetAsync`. Working with `Many` methods has a significant performance gain if you have a lot of items to read or write. The `GetOrAddAsync` method (used in the `GetUserInfoAsync` example in this section) is also defined by ABP Framework to safely read a cache value, fall back to the original data source, and set the cache value in a single method call.

Configuring cache options

`AbpDistributedCacheOptions` is the main options class to configure the caching system. You can configure it in the `ConfigureServices` method of your module class (you can do this in the domain or application layers), as follows:

```
Configure<AbpDistributedCacheOptions>(options =>
{
    options.GlobalCacheEntryOptions
        .AbsoluteExpirationRelativeToNow =
            TimeSpan.FromHours(2);
});
```

I've configured the `GlobalCacheEntryOptions` property to configure the default cache expiration time to 2 hours in this code block.

`AbpDistributedCacheOptions` has some other properties too, as outlined here:

- `KeyPrefix` (`string`; default: `null`): A prefix value that is added to the beginning of all cache keys for that application. This option can be used to isolate your application's cache items when using a distributed cache shared by multiple applications.

- `hideErrors` (`bool`; default: `true`): A value to control the default value of error handling on cache service methods.

As you've seen in the previous examples, these options can be overridden by passing parameters to the methods of the `IDistributedCache` service.

Error handling

When we use an external process (such as Redis) for distributed caching, it is probable to have problems while reading data from and writing data to the cache. The cache server may be offline, or we may have temporary network problems. These temporary problems can be ignored most of the time, especially while trying to read data from the cache. You can safely try to read from the original data source if the cache service is not available at the moment. It may be slower but is better than throwing an exception and failing the current request.

All the `IDistributedCache<T>` methods get an optional `hideErrors` parameter to control the exception-handling behavior. If you pass `false`, then all the exceptions are thrown. If you pass `true`, then ABP hides cache-related errors. If you don't specify a value, the default value is used, as explained in the previous section.

Using the cache in a multi-tenancy application

If your application is multi-tenant, ABP automatically adds the current tenant's ID to the cache key to distinguish between cache values of different tenants. In this way, it provides isolation between tenants.

If you want to create a cache that is shared between tenants, you can use the `[IgnoreMultiTenancy]` attribute for the cache item class, as shown in the following code block:

```
[IgnoreMultiTenancy]
public class MyCacheItem
{ /* ... */ }
```

For this example, `MyCacheItem` values can be accessed by different tenants.

Using Redis as the distributed cache provider

Redis is a popular tool that is used as a distributed cache. ASP.NET Core provides a cache integration package for Redis. You can use it by following Microsoft's documentation (https://docs.microsoft.com/en-us/aspnet/core/performance/caching/distributed), and it works perfectly.

ABP also provides a Redis integration package that extends Microsoft's integration to support the batch operations (such as GetManyAsync, mentioned in the *Using the IDistributedCache<T> interface* section). So, it is suggested to use ABP's integration Volo.Abp.Caching.StackExchangeRedis NuGet package to use Redis as the cache provider. You can install it using the ABP **command-line interface** (**CLI**) with the following command in the directory of the project you want to use:

```
abp add-package Volo.Abp.Caching.StackExchangeRedis
```

After the installation, all you need to do is to add a configuration to the appsettings. json file to connect to the Redis server, as follows:

```
"Redis": {
   "Configuration": "127.0.0.1"
}
```

You write the server address and port (a connection string) to the Configuration option. Please see Microsoft's documentation for details of the configuration: https://docs.microsoft.com/en-us/aspnet/core/performance/caching/distributed.

Invalidating cache values

There is a popular saying that cache invalidation is one of the two hard problems in computer science (the other one is naming things). A cached value is typically a duplication of data originally located somewhere costly to read frequently or a computed value that is costly to recalculate. In such cases, it increases performance and scalability, but the problem begins when the original data changes and makes the cached value outdated. We should carefully watch these changes and remove or refresh the related data in the cache. That is called cache invalidation.

The cache invalidation process depends greatly on the cached data and your application logic. However, there are some specific cases where ABP can help you to invalidate cached data.

One specific case is that we may want to invalidate a cache item when an entity has changed (is updated or deleted). For this case, we can register for events published by ABP Framework. The following code invalidates a user cache item when the related user entity changes:

```
public class MyUserService :
    ILocalEventHandler<EntityChangedEventData<IdentityUser>>,
    ITransientDependency
{
    private readonly IDistributedCache<UserCacheItem>
        _userCache;
    private readonly IRepository<IdentityUser, Guid>
        _userRepository;
    //...omitted other code parts
    public async Task HandleEventAsync(
        EntityChangedEventData<IdentityUser> data)
    {
        await _userCache.RemoveAsync
            (data.Entity.Id.ToString());
    }
}
```

MyUserService registers for an EntityChangedEventData<IdentityUser> local event. This event is triggered when a new IdentityUser entity is created or an existing IdentityUser entity is updated or deleted. The HandleEventAsync method is called in that case with the related entity in the data.Entity property. This method simply removes the user from the cache with the Id value of the changed entity.

Local events work in the current process. That means the handler class (MyUserService here) should be in the same process as the entity change.

About the Event Bus System

Local and distributed events are interesting features of ABP Framework that are not included in this book. See the ABP documentation if you want to learn more about them: https://docs.abp.io/en/abp/latest/Event-Bus.

In this section, we've learned how to work with the distributed caching system, configure options, and deal with error handling. We've also introduced the Redis cache provider installation. Finally, we've introduced automatic ABP events that can help us to invalidate cached values.

The next section will be related to UI localization, the final ABP feature I will introduce in this chapter.

Localizing the user interface

If you are building a global product, you probably want to show the UI localized on the basis of the current user's language. ASP.NET Core provides a system to localize your application's UI. ABP adds some useful features and conventions to make it even easier and flexible.

This section explains how to define the languages that you want to support, create text for different languages, and get the correct text for the current user. You will understand the localization resource concept and embedded localization resource files.

We can begin by defining the languages supported by your application.

Configuring supported languages

The first question about localization is this: *Which languages do you want to support on your UI?* ABP provides a simple configuration to define languages, using `AbpLocalizationOptions`, as shown in the following code block:

```
Configure<AbpLocalizationOptions>(options =>
{
    options.Languages.Add(new LanguageInfo("en", "en",
        "English"));
    options.Languages.Add(new LanguageInfo("tr", "tr",
        "Türkçe"));
    options.Languages.Add(new LanguageInfo("es", "es",
        "Español"));
});
```

You can write that code into the `ConfigureServices` method of your module class. In fact, that configuration is already done (with a lot of languages) when you create a new solution using the ABP application startup templates. You just edit the list as needed.

The `LanguageInfo` constructor takes a few parameters:

- `cultureName`: The culture name (code) for the language, which is set to `CultureInfo.CurrentCulture` on runtime.
- `uiCultureName`: The UI culture name (code) for the language, which is set to `CultureInfo.CurrentUICulture` on runtime.
- `displayName`: Name of the language that is shown to the user while selecting this language. It is suggested to write that name in its original language.
- `flagIcon`: A string value that the UI can use to show a country flag near the language name.

ABP determines one of these languages based on the current HTTP request.

Determining the current language

ABP determines the current language by using the `AbpRequestLocalizationMiddleware` class. This is an ASP.NET Core middleware that is added to the ASP.NET Core request pipeline with the following line of code:

```
app.UseAbpRequestLocalization();
```

When the request passes through this middleware, one of the configured languages is selected and set to `CultureInfo.CurrentCulture` and `CultureInfo.CurrentUICulture`. These are the standard systems of .NET to set and get the current culture in localization.

The current language is selected based on the following HTTP request parameters in the given priority order:

1. If the `culture` query string parameter is set, it is used to determine the current language. An example is `http://localhost:5000/?culture=en-US`.
2. If the `.AspNetCore.Culture` cookie value is set, then it is used as the current language.
3. If the `Accept-Language` HTTP header is set, it is used as the current language. The browser generally sends this last one by default.

About ASP.NET Core's Localization System

The behaviour explained in this section was the default behavior. However, ASP.NET Core's language determination system is more flexible and customizable. Please see Microsoft's documentation for more information: `https://docs.microsoft.com/en-us/aspnet/core/fundamentals/localization.`

After defining the languages we want to support, we can define our localization resources.

Defining a localization resource

ABP is 100% compatible with ASP.NET Core's localization system. So, you can use the `.resx` files as localization resources by following Microsoft's documentation: `https://docs.microsoft.com/en-us/aspnet/core/fundamentals/localization.` However, ABP offers a lightweight, flexible, and extensible way to define localized texts using simple JSON files.

When you create a new solution using the ABP startup templates, the `Domain.Shared` project contains the localization resource of the application with the localization JSON files:

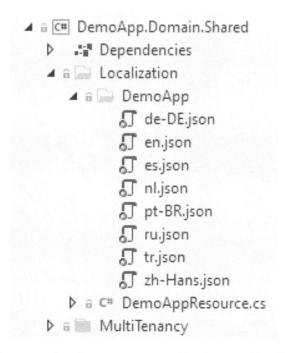

Figure 8.2 – Localization resource and localization JSON files

For this example, the `DemoAppResource` class represents the localization resource. An application can have more than one localization resource, and each defines its own JSON files. You can think of a localization resource as a group of localization texts. It helps to build modular systems where each module has its own localization resource.

A localization resource class is an empty class, as shown in the following code:

```
[LocalizationResourceName("DemoApp")]
public class DemoAppResource
{ }
```

This class refers to the related resource when you want to use texts in that localization resource. The `LocalizationResourceName` attribute sets a string name to the resource. Every localization resource has a unique name that is used in the client-side code to refer to the resource. We will explore client-side localization in *Using localization in the client side* section.

Default Localization Resource of the Application

You typically have a single (default) localization resource in your application that comes with the startup template when creating a new ABP solution. The default localization resource class's name starts with the project name— for example, `ProductManagementResource` if you've specified `ProductManagement` as the project name.

Once we have a localization resource, we can create a JSON file for each language we support.

Working with the localization JSON files

A localization file is a simple JSON-formatted file, as shown in the following code block:

```
{
    "culture": "en",
    "texts": {
        "Home": "Home",
        "WelcomeMessage": "Welcome to the application."
    }
}
```

There are two main root elements in that file, as outlined here:

- `culture`: The culture code for the related language. It matches the culture code that was introduced in *Configuring the supported languages* section.

- `texts`: Contains key-value pairs for the localization texts. The key is used to access the localized texts and should be the same in all JSON files of different languages. The value is the localized text for the current culture (language).

After defining localization texts for each language, we can request localized texts at runtime.

Getting localized texts

ASP.NET Core defines an `IStringLocalizer<T>` interface to get the localized texts in the current culture, where `T` stands for the localization resource class. You can inject that interface into your class, as shown in the following code block:

```
public class LocalizationDemoService : ITransientDependency
{
    private readonly IStringLocalizer<DemoAppResource>
        _localizer;
    public LocalizationDemoService(
        IStringLocalizer<DemoAppResource> localizer)
    {
        _localizer = localizer;
    }
    public string GetWelcomeMessage()
    {
        return _localizer["WelcomeMessage"];
    }
}
```

In the preceding code block, the `LocalizationDemoService` class injects the `IStringLocalizer<DemoAppResource>` service, which is used to access localized texts for the `DemoAppResource` class. In the `GetWelcomeMessage` method, we simply get the localized text for the `WelcomeMessage` key. If the current language is English, it returns `Welcome to the application.`, as we defined in the JSON file in the previous section.

We can pass parameters while localizing text.

Parameterized texts

Localization texts can contain parameters, as shown in the following example:

```
"WelcomeMessageWithName": "Welcome {0} to the application."
```

Parameters can be passed to the localizer, as shown in the following code block:

```
public string GetWelcomeMessage(string name)
{
    return _localizer["WelcomeMessageWithName", name];
}
```

The given name for this example replaces the {0} placeholder.

The fallback logic

The localization system uses fallbacks to parent or default cultures when the requested text is not found in the current culture's JSON file.

For example, assume that you've requested to get a WelcomeMessage text while the current culture (CultureInfo.CurrentUICulture) is de-DE (German–Germany). In that case, one of the following scenarios occurs:

- If you haven't defined a JSON file with "culture": "de-DE" or you have defined a JSON file but it doesn't contain the WelcomeMessage key, then it falls back to the parent culture ("de"), tries to find the given key in that culture, and returns it if available.

- If it is not found in the parent culture, it falls back to the default culture of the localization resource (see the next section to configure the default culture).

- If it is not found in the default culture, then the given key (WelcomeMessage, for this example) is returned as the response.

Configuring localization resources

A localization resource should be added to AbpLocalizationOptions before using it. This configuration is already done in the startup template with the following code:

```
Configure<AbpVirtualFileSystemOptions>(options =>
{
    options.FileSets.AddEmbedded<DemoAppDomainSharedModule>();
});
```

```
Configure<AbpLocalizationOptions>(options =>
{
    options.Resources
        .Add<DemoAppResource>("en")
        .AddBaseTypes(typeof(AbpValidationResource))
        .AddVirtualJson("/Localization/DemoApp");
    options.DefaultResourceType = typeof(DemoAppResource);
});
```

Localization JSON files are generally defined as embedded resources. We are configuring ABP's virtual filesystem (using the `AbpVirtualFileSystemOptions`) to add all embedded files in that assembly into the virtual filesystem so that the localization files are also added.

Then, in the second part, we add `DemoAppResource` to the `Resources` dictionary so that ABP recognizes it. Here, the `"en"` parameter sets the default culture of that localization resource.

ABP's localization system is pretty advanced. It allows you to reuse the texts of a localization resource by inheriting the localization resource from another localization resource. In this example, we are inheriting `AbpValidationResource`, which is defined in ABP Framework and contains standard validation error messages.

The `AddVirtualJson` method is used to set the JSON files related to that resource using the virtual filesystem.

Finally, `DefaultResourceType` sets the default localization resource for that application. You can have a default resource used in some places where you don't specify the localization resource. The next section explains the main usage point of this configuration.

Localizing in special services

Injecting the `IStringLocalizer<T>` service everywhere can be tedious. ABP pre-injects the localizer into some special base classes. When you inherit from these classes, you can directly use the `L` shortcut property to localize texts.

The following example shows how to localize text in an application service method:

```
public class MyAppService : ApplicationService
{
    public async Task FooAsync()
```

```
    {
        var str = L["WelcomeMessage"];
    }
}
```

The L property, in this example, is defined by the ApplicationService base class, so you don't need to inject the IStringLocalizer<T> service manually. You may wonder, as we haven't specified a localization resource, which one is used here. The answer is the DefaultResourceType option that was explained in the previous section.

If you want to specify another localization resource for a particular application service, then set the LocalizationResource property in the constructor of your service:

```
public class MyAppService : ApplicationService
{
    public MyAppService()
    {
        LocalizationResource = typeof(AnotherResource);
    }
    //...
}
```

In addition to the ApplicationService class, some other common base classes, such as AbpController and AbpPageModel, provide the same L property as a shortcut of injecting the IStringLocalizer<T> service.

Using localization on the client side

One of the ABP's benefits for the localization system is that all the localization resources are directly usable on the client-side code.

For example, the following code localizes the WelcomeMessage key in the JavaScript code for an ASP.NET Core MVC/Razor Pages application:

```
var str = abp.localization.localize('WelcomeMessage',
 'DemoApp');
```

DemoApp is the localization resource name, and WelcomeMessage is the localization key here. Client-side localization will be covered in *Part 4, User Interface and API Development,* of this book.

Summary

In this chapter, we learned some essential features that you will need in almost any web application.

The `ICurrentUser` service allows you to get information about the current user in your application. You can work with the standard claims (such as username and ID) and define custom claims based on your requirements.

We have explored the data filtering system that automates filtering data while querying from the database. In this way, we can easily implement some patterns such as soft-delete and multi-tenancy. We also learned how to define custom data filters and disable the filters whenever necessary.

We have understood how the audit log system works to track and save all the operations done by users. We can control the audit log system declaratively and conventionally with attributes and options.

Caching data is another essential concept to improve the performance and scalability of the system. We've learned about ABP's `IDistributedCache<T>` service, which provides a type-safe way to interact with the cache provider and automates some common tasks such as serialization and exception handling.

Finally, we've explored the localization infrastructure of ASP.NET Core and ABP Framework to define and consume localized texts in our applications easily.

Now we have come to the end of this chapter, we've completed *Part 2, Fundamentals of ABP Framework* of this book, covering ABP Framework and ASP.NET Core infrastructure fundamentals. The next part is a practical guide for implementing **domain-driven design (DDD)** using ABP Framework. DDD is one of the core concepts upon which ABP is based. It includes principles, patterns, and practices to build maintainable business solutions.

Part 3: Implementing Domain-Driven Design

This part focuses on **Domain-Driven Design (DDD)**. It starts by introducing DDD overall, then deep dives into the implementation (based on ABP Framework) by showing and explaining explicit rules and examples.

In this part, we include the following chapters:

9
Understanding Domain-Driven Design

The ABP Framework project's main goal is to introduce an architectural approach to application development and provide the necessary infrastructure and tools to implement that architecture with best practices.

Domain-driven design (**DDD**) is one of the core parts of ABP Framework's architecture offering. ABP's startup templates are layered based on DDD principles and patterns. ABP's entity, repository, domain service, domain event, specification, and many other concepts are directly mapped to the tactical patterns of DDD.

Since DDD is a core part of the ABP application development architecture, this book has a dedicated section, *Part 3, Implementing Domain-Driven Design*, for DDD that consists of three chapters. In this book, I will focus on practical implementation details rather than the theoretical and strategic approaches and concepts of DDD. The examples will be mostly based on the *EventHub* project that was introduced in *Chapter 4, Understanding the Reference Solution*. In addition, I will show different examples for some scenarios that the *EventHub* project has no proper examples of.

The next two chapters will show you explicit rules and concrete code examples for implementing DDD to help you learn how to implement DDD with the ABP Framework.

However, in this first chapter, we will look at DDD in general and explore the core technical concepts in the following order:

- Introducing DDD
- Structuring a .NET solution based on DDD
- Dealing with multiple applications
- Understanding the execution flow
- The common principles of DDD

Technical requirements

You can clone or download the source code for the *EventHub* project from GitHub: `https://github.com/volosoft/eventhub`.

If you want to run the solution in your local development environment, you need to have an IDE/editor (such as Visual Studio) to build and run ASP.NET Core solutions. Also, if you want to create ABP solutions, you need to have the ABP CLI installed, as explained in *Chapter 2, Getting Started with ABP Framework*.

Introducing DDD

Before we cover the implementation details, let's define DDD's core concepts and building blocks. Let's begin with the definition of DDD.

What is domain-driven design?

DDD is a software development approach for complex needs where you connect the software's implementation to an evolving model.

DDD is suitable for complex domains and large-scale applications. In the case of simple, short-lived **Create, Read, Update, Delete** (**CRUD**) applications, you typically don't need to follow all the DDD principles. Fortunately, ABP doesn't force you to implement all the DDD principles in every application; you can just use which principles work best for your application. However, following DDD principles and patterns in a complex application helps you build a flexible, modular, and maintainable code base.

DDD focuses on core domain logic rather than the infrastructure details, which are generally isolated from the business code.

Implementing DDD is closely related to **object-oriented programming (OOP)** principles. This book doesn't cover these basic principles, but still, a good understanding of OOP and the **single responsibility, open-closed, Liskov-substitution, interface segregation, and dependency inversion (SOLID)** principles will help you a lot while you're shaping and organizing your code base and implementing DDD in practice.

Now that we've provided this brief definition, we can explore the fundamental layers of DDD.

DDD layers

Layering is a common principle of organizing software solutions to reduce complexity and increase reusability. DDD offers a four-layered model to help you organize your business logic and abstract the infrastructure from the business logic, as shown in the following diagram:

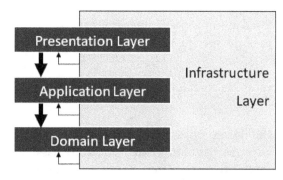

Figure 9.1 – Layers of DDD

The preceding diagram shows the layers and their relationships:

- The **domain layer** contains the essential business objects and implements the core, use case-independent, reusable domain logic of the solution. This layer doesn't depend on any other layer, but all the other layers directly or indirectly depend on it.

- The **application layer** implements the use cases of the applications. A use case is typically an action that's taken by the user through the UI. The **application layer** uses the **domain layer's** objects to perform these use cases.

- The **presentation layer** contains the UI components of the application, such as the views, JavaScript, and CSS files for a web application. It does not directly use the **domain layer** or database objects. Instead, it uses the **application layer**. Typically, for every use case/action that's performed on the UI, there is a corresponding functionality/method on the **application layer**.

- The **infrastructure layer** depends on all the other layers and implements the abstractions that have been defined by these layers. It helps gracefully separate your business logic from third-party libraries and systems, such as database or cache providers.

Each layer of this model has a responsibility and contains various building blocks, which are introduced in the next section.

Building blocks

From a technical perspective, DDD is mostly related to designing your business code by focusing on the domain you are working on. Business logic is separated into two layers – the domain layer and the application layer. The other layers (presentation and infrastructure) are considered as implementation details and should be implemented based on the best practices of the specific technologies you are using, such as Entity Framework.

The domain layer implements the core domain logic by using the following fundamental building blocks:

- **Entity**: An entity is a business object with a state (data) and business logic, both of which work with the properties of this entity. An entity has a unique identifier (ID) that is used to distinguish that entity from the others. This means that two entities with different identifiers are considered different entities, even if all the other properties are identical. For example, the EventHub solution contains the `Event` and `Organization` entities.

- **Value Object**: A value object is another type of business object. Value objects are identified by their state (properties), and they don't have an identifier. This means that two value objects are considered the same if all their properties are the same. Value objects are generally simpler than entities and are typically implemented as immutable. For example, we can create value objects such as address, money, or date.

- **Aggregate** and **Aggregate Root**: An aggregate is a cluster of objects (entities and value objects) that are bound together by an aggregate root object. The aggregate root is responsible for keeping the aggregate objects valid. It implements and coordinates the business logic on these objects. For example, the `Event` entity of the EventHub solution is the aggregate root entity of the Event aggregate, which contains tracks and sessions as sub-collections.

- **Repository**: A repository is a collection-like interface that's used by the domain and application layers to access the persistence system. It hides the complexity of the database provider from the business code.

- **Domain Service**: A domain service is a stateless service (class) that implements core business rules. It is useful to implement domain logic that depends on multiple aggregate types (so that none of these aggregates can be responsible for implementing that logic) or external services. Domain services get/return domain objects and are generally consumed by the application services or other domain services.

- **Specification**: A specification is a named, reusable, testable, and combinable filter that's applied to business objects to select them based on specific business rules.

- **Domain Event**: A domain event is a way of informing other services in a loosely coupled manner when a domain-specific event occurs. It is useful for implementing side effects across multiple aggregates.

The application layer implements the use case of the application by using the following building blocks:

- **Application Service**: An application service is a stateless service (class) that implements the use cases of the application. It typically gets and returns data transfer objects, and its methods are used by the presentation layer. It uses and orchestrates the domain layer objects to perform a specific use case. A use case is typically implemented as a transactional (atomic) process.

- **Data Transfer Objects (DTO)**: A DTO is used to transfer data (state) between the presentation and application layers. It doesn't contain any business logic.

- **Unit of Work (UOW)**: A UOW is a transaction boundary. All the state changes (typically database operations) in a UOW must be implemented as atomic, committed together on success, or rolled back together on failure.

It was important to see the big picture and become familiar with the core building blocks of DDD, which I why I introduced them in brief here. In the next few chapters, we will use them in practice and understand their implementation details. However, in this chapter, I will continue with the big picture and explain how ABP places the layers and building blocks into a .NET solution.

Structuring a .NET solution based on DDD

So far, we have been introduced to the layers and core building blocks of a DDD-based software solution. In this section, we will learn how a .NET solution can be layered based on DDD. I will begin with the simplest possible solution structure. Then, I will explain how ABP's startup solution template evolved into its current structure. Finally, you will understand why the ABP startup solution has that many projects inside it and the purpose of each.

Creating a simple DDD-based .NET solution

Let's start from scratch and keep things simple by creating four projects in our .NET solution, as shown in the following screenshot:

Figure 9.2 – A simple DDD-based .NET solution in Visual Studio

Assuming that we are building a **Customer Relationship Management (CRM)** solution, **Acme** is our company name, and **Crm** is the product name in this example. I've created a separate C# project for each layer. .NET projects perfectly fit into layers as they can physically separate the code base into different packages. A class/type in a project can directly use other classes/types in the same project. However, a class/type can't use a class/type in another project unless you explicitly define the dependency by referencing the other project.

Figure 9.2 shows the projects in the solution in Visual Studio, as well as the dependencies between these projects:

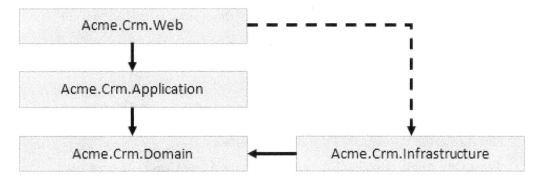

Figure 9.3 – Project dependencies of the simple DDD-based .NET solution

In the preceding diagram, the solid lines represent development-time dependencies (project references), while the dashed line represents runtime dependencies. I will explain the difference later in this section.

To understand these dependencies, we need to know what type of components these projects may contain. We saw which components are located in the domain and application layers in the *Building blocks* section. Here, I will mention some example components that are included in the projects of that CRM solution:

- **Acme.Crm.Domain** is a C# class library project that contains a `Product` class (aggregate root entity) and an `IProductRepository` interface (repository abstraction). The `Product` class represents a product and has some properties such as `Id`, `Name`, and `Price`. `IProductRepository` has some methods to perform database operations for products, such as `Insert`, `Delete`, and `GetList`.

- **Acme.Crm.Infrastructure** is a C# class library project that contains the `CrmDbContext` class (the EF Core data context), which maps the `Product` entity to a database table. It also contains the `EfProductRepository` class, which implements the `IproductRepository` interface.

- **Acme.Crm.Application** is a C# class library project that contains `ProductAppService` (application service), along with some methods to create, update, delete, and get a list of products. This service internally uses the `IProductRepository` interface and the `Product` entity (the domain objects).

- **Acme.Crm.Web** is an ASP.NET Core MVC (Razor Pages) web application. It has a `Products.cshtml` page (and a related JavaScript file) that renders the product data on the UI and allows you to manage (create, edit, and delete) the products. It internally uses `ProductAppService` to perform the actual operations.

Now that we understand the purpose and contents of these projects, let's see why the projects have these dependencies:

- **Acme.Crm.Domain** has no dependencies. In general, the domain layer has a minimal dependency and is abstracted from the infrastructural details.

- **Acme.Crm.Infrastructure** has a dependency on the **Acme.Crm.Domain** project because it needs to access the `Product` class to map it to a database table, and it implements the `IProductRepository` interface.

- **Acme.Crm.Application** has a dependency on the **Acme.Crm.Domain** project because it uses the `IProductRepository` repository and the `Product` entity to perform the use cases.

- Finally, **Acme.Crm.Web** depends on the **Acme.Crm.Application** project because it uses the application service (`ProductAppService`).

The **Acme.Crm.Web** project has one more dependency: it references the **Acme.Crm. Infrastructure** project. It doesn't directly use any class in that project, so there is no need for a direct dependency. However, **Acme.Crm.Web** is also the project that runs the application, and the application needs the infrastructure layer at runtime to use the database. An alternative structure will be discussed in the *Separating the hosting from the UI* section so that you can get rid of that dependency.

This was a minimalistic layering of a DDD-based solution. In the next section, we will use that solution and explain how ABP's startup solution has evolved.

Evolution of the ABP startup solution

ABP's startup solution is more complex than the solution shown in *Figure 9.2*. The following screenshot shows the same solution that was created with the ABP startup template, but this time using the `abp new Acme.Crm` CLI command:

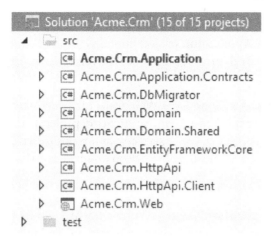

Figure 9.4 – The CRM solution created using the ABP startup template

Let's explain how this solution evolved from the four-project solution explained in the previous section.

Introducing the EntityFrameworkCore project

The minimalistic DDD solution contains the **Acme.Crm.Infrastructure** project, which is assumed to implement all the infrastructural abstractions and integrations. An ABP solution, on the other hand, has a dedicated Entity Framework Core integration project (**Acme.Crm.EntityFrameworkCore**) since we think it is good to create separate projects for such major dependencies, especially for the database integration.

The infrastructure layer can be split into multiple projects. The ABP startup template has no such major dependency. The only infrastructure project is the **Acme.Crm. EntityFrameworkCore** project. If your solution grows, you can create additional infrastructure projects.

With this change, the initial minimalistic DDD-based solution will be as follows:

```
Solution 'Acme.CRM' (4 of 4 projects)
    C#  Acme.Crm.Application
    C#  Acme.Crm.Domain
    C#  Acme.Crm.EntityFrameworkCore
    Acme.Crm.Web
```

Figure 9.5 – Introducing the Entity Framework Core integration project

This change was trivial. It can be thought of as changing the **Acme.Crm.Infrastructure** project's name to **Acme.Crm.EntityFrameworkCore**. The next section will introduce a new project to the solution.

Introducing the application contracts

Currently, the **Acme.Crm.Application** project contains the application service classes. Therefore, the **Acme.Crm.Web** project references the **Acme.Crm.Application** project to use these services.

This design has a problem: the **Acme.Crm.Web** project indirectly references the **Acme.Crm.Domain** project (over the **Acme.Crm.Application** project). This exposes the business objects (such as entities, domain services, and repositories) in the domain layer to the presentation layer and breaks the abstraction and true layering.

The ABP startup template separates the application layer into two projects:

- The **Acme.Crm.Application.Contracts** project, which contains the application service interfaces (such as IProductAppService) and the related DTOs (such as ProductCreationDto).

- The **Acme.Crm.Application** project, which contains the implementations of the application services (such as ProductAppService).

Introducing contracts (interfaces) for the application services has two important advantages:

- The UI layer (the **Acme.Crm.Web** project here) can depend on the service contracts without depending on the implementation, and therefore the domain layer.

- You can share the **Acme.Crm.Application.Contracts** project with a client application to rely on the same service interfaces and reuse the same DTO classes without sharing your business layers.

The EventHub reference solution (introduced in *Chapter 4, Understanding the Reference Solution*) takes advantage of this design and reuses the **Application.Contracts** project between the UI and the HTTP API applications. This way, it easily sets up a tiered architecture where the application layer and the presentation layer are hosted in different applications yet share service contracts.

By separating the application contracts project, the current solution structure will look like the one in the following figure:

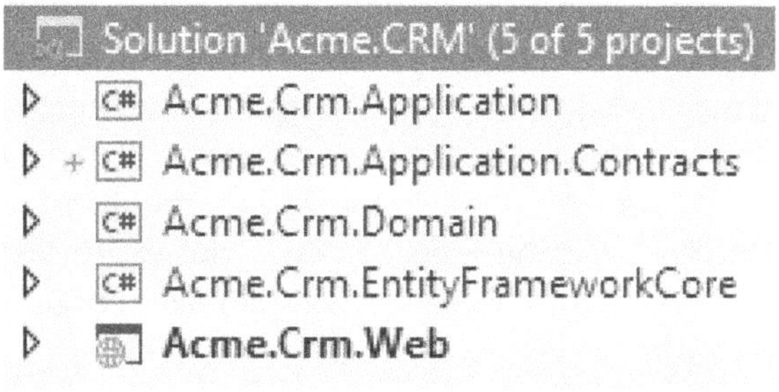

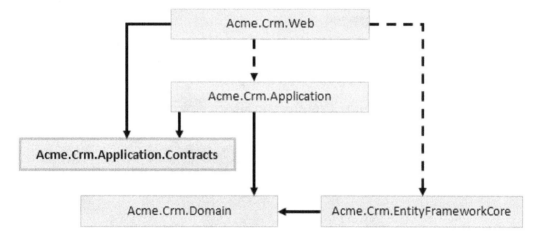

Figure 9.6 – Introducing the application contracts project

With this new design, the project dependency graph will be like in the following figure:

Figure 9.7 – Project dependencies for the application contracts project

The **Acme.Crm.Web** project now only depends on the **Acme.Crm.Application. Contracts** project and should always use the application service interfaces to perform the user interactions.

The **Acme.Crm.Web** project still depends on the **Acme.Crm.Application** and **Acme. Crm.EntityFrameworkCore** projects since we need them at runtime. I have drawn these dependencies with dashed lines to indicate that these project dependencies should not exist in an ideal design, but are necessary for now. I will explain how we can get rid of those dependencies in the *Separating the hosting from the UI* section.

Separating the application contracts from the implementation brings a small problem that we will solve in the next section.

Introducing the domain shared project

Once we have separated the contracts, we can no longer use the objects of the domain layer inside the contracts project because they have no reference to the domain layer, as shown in the previous section. This doesn't seem to be a problem at first glance. We shouldn't use these entities and other business objects in the application service contracts anyway – we should use DTOs instead. However, we still may want to reuse some types or values defined in the domain project.

For example, we may want to reuse a `ProductType` enum in a DTO class or depend on the same constant value for the product name's maximum length. We don't want to duplicate such code parts, but we also can't add a reference to the **Acme.Crm.Domain** project from the **Acme.Crm.Application.Contracts** project. The solution is to introduce a new project to declare such types and values.

We will name this new project **Acme.Crm.Domain.Shared** since this project will be part of the domain layer and shared with the rest of the solution. This project won't contain so many types in practice, but we still don't want to duplicate these types.

With the introduction of the **Acme.Crm.Domain.Shared** project, the new solution structure will be as follows:

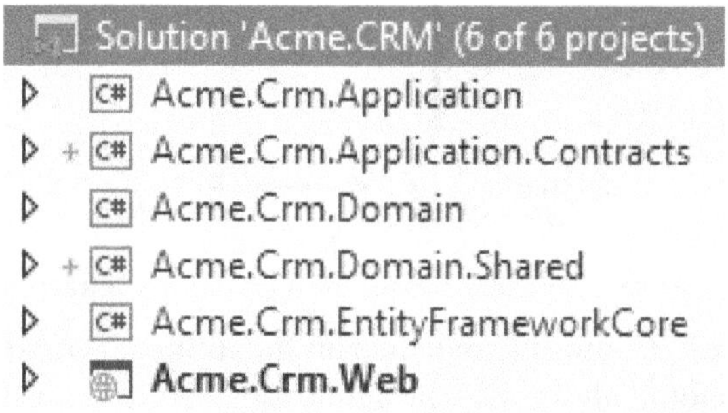

Figure 9.8 – Introducing the domain shared project

The following diagram shows the dependencies between the projects in the solution:

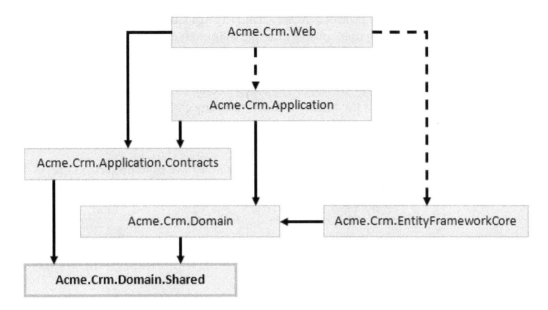

Figure 9.9 – Project dependencies for the domain shared project

The new **Acme.Crm.Domain.Shared** project is used by the **Acme.Crm.Domain** and **Acme.Crm.Application.Contracts** projects. In this way, directly or indirectly, all the other projects in the solution can use the types in that new project.

At this point, the fundamental layers of the ABP startup solution are complete. However, if you look at *Figure 9.4*, you will see that the ABP startup solution has three more projects. We will discuss these in the following subsections.

Introducing the HTTP API layer

In *Figure 9.4*, you can see that the ABP startup solution has two HTTP-related projects.

First, the **Acme.Crm.HttpApi** project contains the API Controllers (that is, the REST APIs) of the solution. This project was introduced with the idea that separating the API from the UI would be better to organize and develop the solution.

Separating the HTTP API layer as a class library project makes some advanced scenarios possible by allowing them to be reused. The EventHub solution takes advantage of this separation by using the HTTP API layer as a proxy in the UI layer (the UI and HTTP API are hosted in different applications in that solution). See the *Main website* and *Main HTTP API* sections of *Chapter 4, Understanding the Reference Solution*, to learn how it works.

The second HTTP API-related project is **Acme.Crm.HttpApi.Client**. This is a class library project that is not being used for this example solution but can be used in more advanced scenarios. You can use this library from a client application (it can be your application or a third-party .NET client) to easily consume your HTTP APIs. It uses ABP's Dynamic C# Client Proxy system, as will be explained in *Chapter 14, Building HTTP APIs and Real-Time Services*. Most of the time, you don't make any changes to this project, but it *automagically* works. The EventHub solution uses this technique to perform HTTP API requests from the UI application.

By adding two new projects for the HTTP API layer, we now have eight projects in the solution, as shown in the following screenshot:

> ⌐ Solution 'Acme.CRM' (8 of 8 projects)
> ▷   C# Acme.Crm.Application
> ▷   C# Acme.Crm.Application.Contracts
> ▷   C# Acme.Crm.Domain
> ▷   C# Acme.Crm.Domain.Shared
> ▷   C# Acme.Crm.EntityFrameworkCore
> ▷ + C# Acme.Crm.HttpApi
> ▷ + C# Acme.Crm.HttpApi.Client
> ▷   ⊕ **Acme.Crm.Web**

Figure 9.10 – Adding the HTTP API projects to the solution

The following diagram shows the new dependency graph after adding these new projects (this time, I've removed the Acme.Crm. prefix from the project names to make them fit into the diagram):

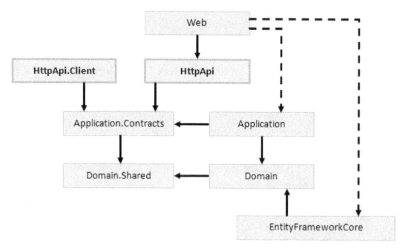

Figure 9.11 – Project dependencies for the HTTP API layer

The **Acme.Crm.HttpApi** and **Acme.Crm.HttpApi.Client** projects depend on the **Acme.Crm.Application.Contracts** project because the server and client share the same contracts (application service interfaces). The **Acme.Crm.Web** project depends on the **Acme.Crm.HttpApi** project since it serves the APIs at runtime. This example solution has a single application at runtime. You can revisit the EventHub solution structure that was provided in *Chapter 4, Understanding the Reference Solution*, to see these projects in a more complex environment with multiple applications at runtime.

> **Discarding the HTTP API Layer**
>
> Not every application needs to have HTTP APIs (that is, REST APIs). In this case, you can even remove this project from the solution. Also, if you like, you can move your API controllers to the **Acme.Crm.Web** project and discard the **Acme.Crm.HttpApi** project.

The next section will explain the last project in the solution.

Understanding the database migrator project

In *Figure 9.4*, there is one more project named **Acme.Crm.DbMigrator**. This is a console application that can be used to apply EF Core code-first migrations to the database. It is a utility application and not part of the essential solution, so there is no need to investigate its details here.

> **Test Projects in the Solution**
>
> Besides these nine projects, there are six more projects in the solution under the test folder. They are unit/integration tests projects separately configured for each layer. One of them (**Acme.Crm.HttpApi.Client.ConsoleTestApp**) demonstrates how to consume HTTP APIs using the **Acme.Crm.HttpApi.Client** project. You can explore them yourself.

These are all the projects in the ABP startup solution. The solution structure that's been provided is the architectural model, followed by all the pre-built official ABP application modules. This model makes it possible to reuse the application modules in various scenarios, thanks to its flexibility and modularity.

In the next section, we will discuss an additional project that can be used to separate the hosting from the UI application.

Separating the hosting from the UI

One annoying thing in the architectural model shown in *Figure 9.11* is that the **Web** project references the **Application** and **EntityFramework** projects. None of the pages/ classes in the **Web** project directly use classes in these projects. However, since the **Web** project is the project that runs the application, we needed to reference these projects to make them available at runtime.

This structure is not a big problem, so long as you do not accidentally leak your domain and database layer objects into the presentation (web) layer. However, if you are worried and do not want to set development time dependencies for these runtime dependencies, you can add one more project, **Acme.Crm.Web.Host**, as shown in the following screenshot:

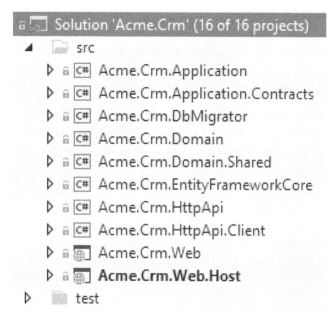

Figure 9.12 – Adding a separate hosting project

With this change, the **Acme.Crm.Web** project becomes a class library project rather than a final application. It only contains the presentation layer pages/components of the application; it does not contain the Startup.cs, Program.cs, and appsettings. json files. The **Acme.Crm.Web.Host** project becomes responsible for hosting by bringing all the projects together at runtime. It doesn't contain any application UI page or component.

I think this design is better. It gracefully extracts the hosting configuration details from the UI layer, removes the runtime dependencies, and keeps it more focused. However, we haven't separated the hosting application in the ABP startup template since most of the developers already find the ABP startup template complicated (compared to single-project ASP.NET Core startup templates). This is because there are many projects inside it, and we didn't want to add one more. I believe that a solution with multiple projects, and with less code in each project, is a better approach than a single project with everything in one place.

You can find the solution with a separate host project in this book's GitHub repository at `https://github.com/PacktPublishing/Mastering-ABP-Framework/tree/main/Samples/Chapter-09/SeparateHosting` and explore the structure provided.

In this section, you understood the roles that each project has in the ABP startup template, so you should be more comfortable while developing your solutions. In the next section, we will briefly revisit the EventHub reference solution from a DDD perspective.

Dealing with multiple applications

So, we've learned the purpose of each of the projects in the ABP startup solution. It is a good starting point for a well-architected software solution. It sets up the layers properly, with a single domain layer and a single application layer (which is used by a single web application). However, in the real world, software solutions may be more complex. You may have multiple applications (on the same system) or may need to separate your domain into multiple sub-domains to reduce the complexity of each sub-domain.

DDD addresses the design of complex software solutions. One of the main purposes of separating the business logic into domain logic and application logic is to correctly organize your code base when there are multiple applications in your solution. When you have multiple applications, you have multiple application layers. Each of these layers implements the application-specific business logic of the related application, yet still shares the same core domain logic by using the same domain layer.

The *EventHub* project (introduced in *Chapter 4, Understanding the Reference Solution*) has two web applications. One of these applications is the main website that is used by end users. The other one is the admin (back office) application, which is used by system administrators. These applications have different user interfaces, different use cases, different authorization rules, and different performance, localization, caching, and scaling requirements. Separating these differences into two application layers helps us isolate these application-specific business and infrastructure requirements from each other. These applications share the core business logic that we don't want to duplicate across the applications. This means that two application layers use the same domain layer, as shown in the following diagram:

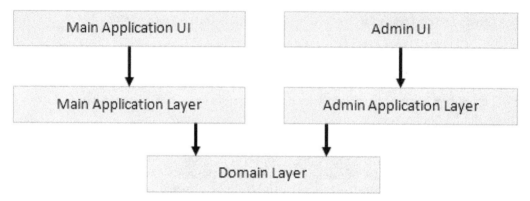

Figure 9.13 – EventHub – multiple application layers and a single domain layer

When we have multiple applications, separating the business logic between the application and domain layers becomes even more important. Leaking domain logic into the application layers duplicates it. On the other hand, placing application-specific logic in the domain layer leads you to coupling the business logic of different applications and writing many conditional statements to make the domain layer usable by these applications. Both of these situations make your code base buggy and difficult to maintain.

Domain logic versus application logic separation is important. We will return to this topic in *Chapter 11, DDD – The Application Layer*, after understanding the domain layer and the application layer building blocks. But before that, let's continue with the big picture and learn how a web request is executed in a DDD-based application.

Understanding the execution flow

We've introduced many building blocks and their descriptions, as well as how these building blocks are placed in layers in a .NET solution. In this section, we will explore how an HTTP request is executed in a typical web application that has been layered based on DDD. The following diagram shows the layers in action:

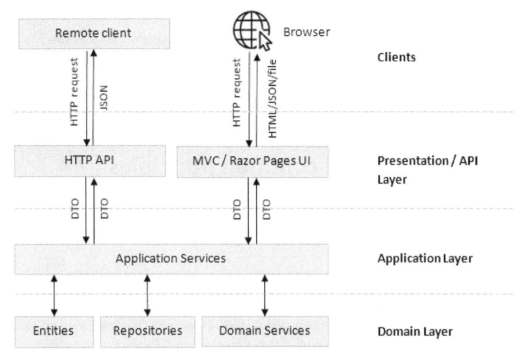

Figure 9.14 – Execution flow through the layers

A request starts with a request from a client application. The client can be a browser that expects an HTML page (with its CSS/JavaScript files) or a data result (such as JSON). In this case, a Razor Page can process the request and returns an HTML page. If the application making the request is another kind of client (such as a console application), you probably respond to the request from an HTTP API (an API controller) endpoint and return a plain data result.

The MVC page (in the presentation layer) processes the UI logic, may perform some data conversions, and delegates the actual operation to a method of an application in the application layer. The application service may take a DTO, implement the use case logic, and return a resulting DTO to the presentation layer.

The application service internally uses the domain objects (entities, repositories, domain services, and more) to coordinate the business operation. The business operation should be a unit of work. This means it should be atomic. All the database operations in a use case (typically, an application method) should be committed or rolled back together.

The presentation and application layers typically implement the cross-cutting concerns, such as authorization, validation, exception handling, caching, audit logging, and so on.

As you learned in the previous chapters, ABP Framework provides a complete infrastructure for all these cross-cutting concerns and automates them wherever possible. It also provides proper base classes and practical conventions to help you structure your business components and implement DDD with best practices.

As the last part of this chapter, we will see some common principles of DDD in the next section.

Understanding the common principles

DDD focuses on how you design your business code. It cares about state changes and how the business objects interact – how to create an entity, how to change its properties by applying (and even forcing) the business rules and constraints, and how to preserve the data validity and integrity.

DDD doesn't care about reporting or mass querying. You may take the power of a reporting tool to create cool dashboards for your application. You can fully use your underlying database provider's features for high performance. You can even duplicate the data in another database provider for read-only reporting purposes. You are free to do anything, so long as you don't mix the infrastructure details with your business code. All these are the concerns we should care about as a developer, but DDD doesn't care.

DDD also doesn't care about the infrastructure details; you are expected to isolate your business code from these details with proper abstractions. Two of these abstractions are especially important since they take a big place in your code base: the presentation technology and the database provider. In the next few sections, I will explain these two principles and discuss if we need to implement them.

Database provider independence

It is a good practice to abstract the database integration in a DDD-based software solution. Your domain and application layers should be database and even ORM independent, in theory. There are some good reasons behind this suggestion. If you implement it, the following will occur:

- Your database provider (ORM or DBMS) may change in the future without affecting your business code. This makes your business code longer-lived.

- Your domain and application layers become more focused on your business code by hiding the data access logic behind the repositories.

- You can mock the database layer for automated tests more efficiently.

The ABP startup template follows this principle – it doesn't include references to the database provider from the domain and application layers. ABP Framework already provides the infrastructure to implement the repository pattern easily. The ABP startup template also comes with the database layer, which uses an in-memory database instance for automated tests.

The last two of these reasons are important and easy to apply with ABP Framework. However, the first one is not so easy. In the beginning, it seems like you make your business code ORM/database independent when you place your data access logic behind the repositories. However, it is not that simple. Let's assume that you are currently using EF Core with SQL Server (a relational database) and want to design your business code and entities so that you can easily switch to MongoDB (a document database) later. If you want to accomplish that, you must take the following into account:

- You can't assume that you have the change tracking system of EF Core because the MongoDB .NET driver doesn't provide that feature. So, you should always manually update the changed entities at the end of your business logic.

- You can't add navigation or collection properties to your entity where these properties are types of other aggregates. You must strictly implement the aggregate pattern (as will be explained in *Chapter 10, DDD – The Domain Layer*) by respecting the aggregate boundaries. This restriction deeply affects your entity design and the business code that works on your entities.

As you can see, being database-agnostic requires care when it comes to designing the entity and affects your code base.

You may be wondering, do you need it? Will you change the database provider in the future? If you change it later, how much effort do you need to make regarding that change? Is it more than your current effort to make it database-independent? Even if you try to do it, will it be truly database-independent (you may not know it before trying to switch)?

All ABP pre-built application modules are designed to be independent of the database provider, and the same business code works both on EF Core and MongoDB. This is necessary since they are reusable modules and can't assume a database provider. On the other hand, a final application can make this assumption. I still suggest hiding the data access code behind the repositories, and ABP makes this very easy. However, if you want to go with an EF Core dependency, I see no problem with that.

Presentation technology-agnostic

UI frameworks are the most dynamic systems in the software industry. There are plenty of alternatives, and the trending approaches and tools are rapidly changing. Coupling your business code with your UI code would be a bad idea.

Implementing this principle is more important and relatively easier, especially with ABP Framework. The ABP startup template comes with proper layering. ABP Framework provides many abstractions that you can use in your application and domain layers without depending on ASP.NET Core or any other UI framework.

Summary

In this first chapter on DDD, we looked at the four fundamental layers and the core building blocks in these layers. The ABP startup template is more complex than that four-layered structure. You learned how the startup template has evolved by one change at a time, and you understood the reasons behind these changes.

Regarding DDD, you learned that the business logic is separated into two layers: the application layer and the domain layer. We discussed how to deal with multiple applications that share the same domain logic by referencing the EventHub example solution.

We then understood how an HTTP request is executed and passed through the layers in a typical DDD-based software. Finally, we discussed isolating your application and domain layers from the infrastructure details, especially the database providers and UI frameworks.

This chapter aimed to show the big picture and the fundamental concepts of DDD. The next chapter will focus on implementing domain layer building blocks, such as aggregates, repositories, and domain services.

10
DDD – The Domain Layer

The previous chapter was an overall view of **Domain-Driven Design** (DDD), where you learned about the fundamental layers, building blocks, and principles of DDD. You also gained an understanding of the structure of the ABP solution and its relation to DDD.

This chapter completely focuses on the implementation details of the domain layer with a lot of code examples and best practice suggestions. Here are the topics we will cover in this chapter:

- Exploring the example domain
- Designing aggregates and entities
- Implementing domain services
- Implementing repositories
- Building specifications
- Publishing domain events

Technical requirements

You can clone or download the source code of the *EventHub* project from GitHub: `https://github.com/volosoft/eventhub`.

If you want to run the solution in your local development environment, you need to have an IDE/editor (such as Visual Studio) to build and run ASP.NET Core solutions. Also, if you want to create ABP solutions, you need to have the ABP CLI installed, as explained in *Chapter 2, Getting Started with ABP Framework*.

Exploring the example domain

The examples in this chapter and the next chapter will mostly be based on the EventHub solution. So, it is essential to understand the domain first. *Chapter 4, Understanding the Reference Solution*, has already explained the solution. You can check it if you want to refamiliarize yourself with the application and the solution structure. Here, we will explore the technical details and the domain objects.

The following list introduces and explains the main concepts of the domain:

- **Event** is the root object that represents an online or in-person event. An event has a title, description, start time, end time, registration capacity (optional), and a language (optional) as the main properties.

- An event is created (organized) by an **Organization**. Any **User** in the application can create an organization and organize events within that organization.

- An event can have zero or more **Tracks** with a track name (typically a simple label such as 1, 2, 3 or A, B, C). A track is a sequence of sessions. An event with multiple tracks makes it possible to organize parallel sessions.

- A track contains one or more **Sessions**. A session is a part of the event where attendees typically listen to a speaker for a certain length of time.

- Finally, a session can have zero or more **Speakers**. A speaker is a person who talks in the session and makes a presentation. Generally, every session will have a speaker. But sometimes, there can be multiple speakers, or there can be no speaker associated with the session. *Figure 10.1* (in the next section) shows the relation of an event to its tracks, sessions, and speakers.

- Any user in the application can **Register** for an event. Registered users are notified before the event starts or if the event time changes.

You've learned about the fundamental objects in the EventHub application. The next section explains the first building block of DDD: aggregates.

Designing aggregates and entities

It is very important to design your entities and aggregate boundaries since the rest of the solution components will be based on that design. In this section, we will first understand what an aggregate is. Then we will see some key principles of an aggregate design. Finally, I will introduce some explicit rules and code examples to understand how we should implement aggregates.

What is an aggregate root?

An aggregate is a cluster of objects bound together by an aggregate root object. The aggregate root object is responsible for implementing the business rules and constraints related to the aggregate, keeping the aggregate objects in a valid state and preserving the data integrity. The aggregate root and the related objects have methods to achieve that responsibility.

The **Event** aggregate shown in the following figure is a good example of aggregates:

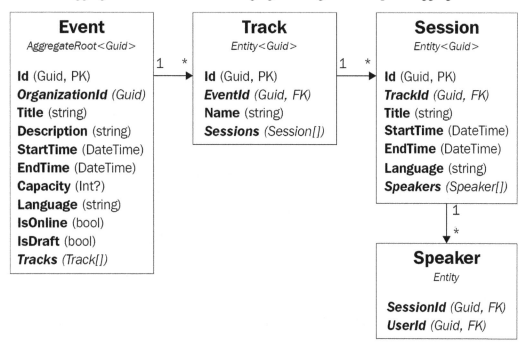

Figure 10.1 – The Event aggregate

The examples in this chapter will mostly be based on the `Event` aggregate since it represents the essential concept of the EventHub solution. So, we should understand its design:

- The `Event` object is the aggregate root here, with a `GUID` primary key. It has a collection of `Track` objects (an event can have zero or more tracks).

- A `Track` is an entity with a `GUID` primary key and contains a list of `Session` objects (a track should have one or more sessions).

- A `Session` is also an entity with a `GUID` primary key and contains a list of `Speaker` objects (a session can have zero or more speakers).

- A `Speaker` is an entity with a composite primary key that consists of `SessionId` and `UserId`.

`Event` is a relatively complex aggregate. Most of the aggregates in an application will consist of a single entity, the aggregate root entity.

The aggregate root is also an entity with a special role in the aggregate: it is the root entity of the aggregate and is responsible for sub-collections. I will refer to the term **Entity** for both aggregate root and sub-collection entities. So, the entity rules are valid for both object types, unless I specifically refer to one of them.

In the next sections, I will introduce two fundamental properties of an aggregate, a single unit and a serialized object.

A single unit

An aggregate is retrieved (from the database) and stored (in the database) as a single unit, including all the properties and sub-collection entities. For example, if you want to add a new `Session` to an `Event`, you should do the following:

1. Read the related `Event` object from the database with all the `Track`, `Session`, and `Speaker` objects.

2. Use a method of the `Event` class to add the new `Session` object to a `Track` of the `Event`.

3. Save the `Event` aggregate to the database together with the new changes.

This might seem inefficient to developers who are used to working with relational databases and ORMs such as EF Core. However, it is necessary because that's how we keep the aggregate objects consistent and valid by implementing the business rules.

Here is a simplified example application service method implementing that process:

```
public class EventAppService
    : EventHubAppService, IEventAppService
{
    //...
    public async Task AddSessionAsync(Guid eventId,
                                      AddSessionDto input)
    {
        var @event =
            await _eventRepository.GetAsync(eventId);
        @event.AddSession(input.TrackId, input.Title,
            input.StartTime, input.EndTime);
        await _eventRepository.UpdateAsync(@event);
    }
}
```

For this example, the event.AddSession method internally checks whether the new session's start time and end time conflict with another session on the same track. Also, the time range of the session should not overflow the event time range. We may have other business rules, too. We may want to limit the number of sessions in an event or check whether the session's speaker has another talk in the same time range.

Remember that DDD is for state changes. If you need a mass query or to prepare a report, you can optimize your database query as much as possible. However, for any change on an aggregate, we need all the objects on that aggregate to apply the business rules related to that change. If you are worried about the performance, see the *Keep aggregates small* section.

At the end of the method, we've updated the Event entity using the repository's UpdateAsync method. If you are working with EF Core, you do not need to explicitly call the UpdateAsync method, thanks to EF Core's change tracking system. The changes will be saved since ABP's Unit of Work system calls the DbContext. SaveChangesAsync() method for you. However, for example, the MongoDB .NET Driver has no change tracking system, and you should explicitly call the UpdateAsync method to the Event object if you're using MongoDB.

> **About the IRepository.GetAsync Method**
>
> The `GetAsync` method of the repository (used in the preceding example code block) retrieves the `Event` object as an aggregate (with all sub-collections) as a single unit. It works out of the box for MongoDB, but you need to configure your aggregate for EF Core to enable that behavior. See the *The Aggregate pattern* section in *Chapter 6, Working with the Data Access Infrastructure*, to remember how to configure it.

Retrieving and saving an aggregate as a single unit brings us the opportunity to make multiple changes to the objects of a single aggregate and save all of them together with a single database operation. This way, all the changes in that aggregate become atomic by nature without needing an explicit database transaction.

> **The Unit of Work System**
>
> If you need to change multiple aggregates (of the same or different types), you still need a database transaction. In this case, ABP's Unit of Work system (explained in *Chapter 6, Working with the Data Access Infrastructure*) automatically handles database transactions by convention.

A serializable object

An aggregate should be serializable and transferrable on the wire as a single unit, including all of its properties and sub-collections. That means you are able to convert it to a byte array or an XML or JSON value, then deserialize (re-construct) it from the serialized value.

EF Core does not serialize your entities, but a document database, such as MongoDB, may serialize your aggregate to a BSON/JSON value to store in the data source.

This principle is not a design requirement for an aggregate, but it is a good guide while determining the aggregate boundary. For example, you cannot have properties referencing entities of other aggregates. Otherwise, the referenced object is also serialized as a part of your aggregate.

Let's look at some more principles. The first rule, introduced in the next section, is the key practice to make an aggregate serializable.

Referencing other aggregates by their ID

The first rule says that an aggregate (including the aggregate root and other classes) should not have navigation properties to other aggregates but can store their ID values when necessary.

This rule makes the aggregate a self-contained, serializable unit. It also helps to not leak the business logic of an aggregate into another aggregate, by hiding aggregate details from other aggregates.

See the following example code block:

```
public class Event : FullAuditedAggregateRoot<Guid>
{
    public Organization Organization { get; private set; }
    public string Title { get; private set; }
    ...
}
```

The Event class has a navigation property to the Organization aggregate, which is prohibited by this rule. If we serialize an Event object to a JSON value, the related Organization object is also serialized.

In a proper implementation, the Event class can have an OrganizationId property for the related Organization:

```
public class Event : FullAuditedAggregateRoot<Guid>
{
    public Guid OrganizationId { get; private set; }
    public string Title { get; private set; }
    ...
}
```

Once we have an Event object and need to access the related organization details, we should query the Organization object from the database using OrganizationId (or perform a JOIN query to load them together at the beginning).

If you are using a document database, such as MongoDB, this rule will seem natural to you. Because if you add a navigation property to the `Organization` aggregate, then the related `Organization` object is serialized and saved in the collection of `Event` objects in the database, which duplicates the organization data and copies it into all events. However, with relational databases, ORMs such as EF Core allow you to use such navigation properties and handle the relation without any problem. I still suggest implementing this rule since it keeps your aggregates simpler and reduces the complexity of loading related data. If you don't want to apply this rule, you can refer to the *Database provider independence* section of *Chapter 9, Understanding Domain-Driven Design*.

The next section expresses a best practice: keep your aggregates small!

Keep aggregates small

Once we load and save an aggregate as a single unit, we may have performance and memory usage problems if the aggregate is too big. Keeping an aggregate simple and small is an essential principle, not just for performance but also to reduce the complexity.

The main aspect that makes an aggregate big is the potential number of sub-collection entities. If a sub-collection of an aggregate root contains hundreds of items, that's a sign of a bad design. In a good aggregate design, the items in a sub-collection should not exceed a few dozen and should remain below 100–150 in edge cases.

See the `Event` aggregate in the following code block:

```
public class Event : FullAuditedAggregateRoot<Guid>
{
    ...
  public ICollection<Track> Tracks { get; set; }
  public ICollection<EventRegistration> Registrations {
      get; set; }
}

public class EventRegistration : Entity
{
    public Guid EventId { get; set; }
    public Guid UserId { get; set; }
}
```

There are two sub-collections of the `Event` aggregate in this example: `Tracks` and `Registrations`.

The `Tracks` sub-collection is a collection of parallel tracks in the event. It typically contains a few items, so there's no problem with loading the tracks while loading an `Event` entity.

The `Registrations` sub-collection is a collection of the registration records for the event. Thousands of people can register for a single event, which will be a significant performance problem if we load all registered people whenever loading an event. Also, most of the time, we don't need all registered users while manipulating an `Event` object. So, it would be better not to include the collection of registered people in the `Event` aggregate. In this example, the `EventRegistration` class is a sub-collection entity. For a better design, we should make it a separate aggregate root class.

There are three main considerations while determining the boundaries of an aggregate:

- The objects related and used together
- Data integrity, validity, and consistency
- The load and save performance of the aggregate (as a technical consideration)

In real life, most aggregate roots won't have any sub-collections. When you think to add a sub-collection to an aggregate, think about the object size as a technical factor.

> **Concurrency Control**
>
> Another problem with big aggregate objects is that they increase the probability of the concurrent update problem since big objects are more likely to be changed by multiple users simultaneously. ABP Framework provides a standard model for concurrency control. Please refer to the documentation: `https://docs.abp.io/en/abp/latest/Concurrency-Check`.

In the next section, we will discuss single and composite primary keys for entities.

Determining the primary keys for entities

Entities are determined by their ID (a unique identifier or the **primary key (PK)**) rather than by other properties. ABP Framework allows you to choose any type of PK that your database provider supports. However, it uses the `Guid` type for the entities of the pre-built modules. It also assumes that the user ID and tenant ID types are `Guid`. We discussed this topic in the *The GUID PK* section of *Chapter 6, Working with the Data Access Infrastructure*.

ABP also allows you to use composite primary keys for your entities. A composite primary key consists of two or more properties (of an entity) that become a unique value together.

As a best practice, use a single primary key (a `Guid` value, an incremental integer value, or whatever you want) for the aggregate root. You can use a single or composite primary for sub-collection entities.

> **Composite Keys in Non-Relational Databases**
>
> Composite primary keys for sub-collection entities are generally used in relational databases because sub-collections have their own tables in a relational database. However, in a document database (such as MongoDB), you don't define primary keys for sub-collection entities since they don't have their own database collections. Instead, they are stored as a part of the aggregate root.

In the *EventHub* project, `Event` is an aggregate root with a `Guid` primary key. `Track`, `Session`, and `Speaker` are sub-collection entities as a part of the `Event` aggregate. `Track` and `Session` entities have `Guid` primary keys, but the `Speaker` entity has a composite primary key.

The `Speaker` entity class is shown in the following code block:

```
public class Speaker : Entity
{
    public Guid SessionId { get; private set; }
    public Guid UserId { get; private set; }

    public Speaker(Guid sessionId, Guid userId)
    {
        SessionId = sessionId;
        UserId = userId;
    }

    public override object[] GetKeys()
    {
        return new object[] {SessionId, UserId};
    }
}
```

`SessionId` and `UserId` compose the unique identifier for the `Speaker` entity. The `Speaker` class is derived from the `Entity` class (without a generic argument). When you derive from the non-generic `Entity` class, ABP Framework forces you to define the `GetKeys` method to obtain the components of the composite key. If you want to use composite keys, refer to the documentation of your database provider (such as EF Core) to learn how to configure them.

Beginning with the next section, we will look at the implementation details of aggregates and entities.

Implementing entity constructors

A constructor method is used to create an object. The compiler creates a default parameterless constructor when we don't explicitly add a constructor to the class. Defining a constructor is a good way to ensure an object is properly created.

An entity's constructor is responsible for creating a valid entity. It should get the required values as the constructor parameters to force us to supply these values during object creation so the new object is useful just after creation. It should check (validate) these parameters and set the properties of the entity. It should also initialize sub-collections and perform additional initialization logic if necessary.

The following code block shows an entity (an aggregate root entity) from the *EventHub* project:

```
public class Country : BasicAggregateRoot<Guid>
{
    public string Name { get; private set; }

    private Country() { } // parameterless constructor

    public Country(Guid id, string name)
        //primary constructor
        : base(id)
    {
        Name = Check.NotNullOrWhiteSpace(
            name, nameof(name),
            CountryConsts.MaxNameLength);
    }
}
```

Country is a very simple entity that has a single property: Name. The Name property is required, so the Country primary constructor (the actual constructor that is intended to be used by the application developers) forces the developer to set a valid value to that property by defining a name parameter and checking whether it is empty or exceeds a maximum length constraint. Check is a static class of ABP Framework with various methods used to validate method parameters and throw an ArgumentException error for invalid parameters.

The Name property has a private setter, so there is no way to change this value after the object creation. We can assume that countries don't change their names, for this example.

The Country class's primary constructor takes another parameter, Guid id. We don't use Guid.NewGuid() in the constructor since we want to use the IGuidGenerator service of ABP Framework, which generates sequential GUID values (see *The GUID PK* section of *Chapter 6, Working with the Data Access Infrastructure*). We directly pass the id value to the base class (BasicAggregateRoot<Guid> in this example) constructor, which internally sets the Id property of the entity.

> **The Need for Parameterless Constructors**
>
> The Country class also defines a private, parameterless constructor. This constructor is just for ORMs, so they can construct an object while reading from the database. Application developers do not use it.

Let's see a more complex example, showing the primary constructor of the Event entity:

```
internal Event(
    Guid id,
    Guid organizationId,
    string urlCode,
    string title,
    DateTime startTime,
    DateTime endTime,
    string description)
    : base(id)
{
    OrganizationId = organizationId;
    UrlCode = Check.NotNullOrWhiteSpace(urlCode, urlCode,
            EventConsts.UrlCodeLength,
            EventConsts.UrlCodeLength);
```

```
    SetTitle(title);
    SetDescription(description);
    SetTimeInternal(startTime, endTime);
    Tracks = new Collection<Track>();
}
```

The `Event` class's constructor takes the minimal required properties as the parameters, and checks and sets them to the properties. All these properties have private setters (see the source code) and are set via the constructor or some methods of the `Event` class. The constructor uses these methods to set the `Title`, `Description`, `StartTime`, and `EndTime` properties.

Let's see the `SetTitle` method's implementation:

```
public Event SetTitle(string title)
{
    Title = Check.NotNullOrWhiteSpace(title, nameof(title),
            EventConsts.MaxTitleLength,
            EventConsts.MinTitleLength);
    Url = EventUrlHelper.ConvertTitleToUrlPart(Title) + "-"
            + UrlCode;
    return this;
}
```

The `SetTitle` method assigns the given `title` value to the `Title` property by checking the constraints. It then sets the `Url` property, a calculated value based on the `Title` property, and the `UrlCode` property. This method is `public`, to use later when we need to change the `Event` entity's `Title` property.

`UrlCode` is an eight-character random unique value that is sent to the constructor and never changes. Let's see another method that the constructor calls:

```
private Event SetTimeInternal(DateTime startTime,
                              DateTime endTime)
{
    if (startTime > endTime)
    {
        throw new BusinessException(EventHubErrorCodes
            .EventEndTimeCantBeEarlierThanStartTime);
    }
```

```
        StartTime = startTime;
        EndTime = endTime;
        return this;
    }
```

Here, we have a business rule: the `StartTime` value cannot be later than the `EndTime` value.

The `EventHub` constructor is `internal` to prevent creating `Event` objects out of the domain layer. The application layer should always use the `EventManager` domain service to create a new `Event` entity. In the next section, we will see why we've designed it like that.

Using services to create aggregates

The best way to create and initialize a new entity is using its public constructor because it is the simplest way. However, in some cases, creating an object requires some more complex business logic that is not possible to implement in the constructor. For such cases, we can use a factory method on a domain service to create the object.

The `Event` class's primary constructor is `internal`, so the upper layers cannot directly create a new `Event` object. We should use the `EventManager` `CreateAsync` method to create a new `Event` object:

```
public class EventManager : DomainService
{
    ...
    public async Task<Event> CreateAsync(
        Organization organization,
        string title,
        DateTime startTime,
        DateTime endTime,
        string description)
    {
        return new Event(
            GuidGenerator.Create(),
            organization.Id,
            await _eventUrlCodeGenerator.GenerateAsync(),
            title,
```

```
        startTime,
        endTime,
        description
    );
}
}
```

We will return to domain services later, in the *Implementing domain services* section of this chapter. With this simple `CreateAsync` method, we are creating a valid `Event` object and returning the new object. We needed such a factory method because we used the `eventUrlCodeGenerator` service to generate URL code for the new event. The `eventUrlCodeGenerator` service internally creates a random, eight-character code for the new event and also checks whether that code was used by another event before (see its source code if you want to learn more). That's why it is `async`: it performs a database operation.

We've used a factory method of a domain service to create a new `Event` object because the `Event` class's constructor cannot use the `eventUrlCodeGenerator` service. So, you can create factory methods if you need external services/objects while creating a new entity.

> **Factory Service versus Domain Service**
>
> An alternative approach can be creating a dedicated class for the factory method. That means we could create an `EventFactory` class and move the `CreateAsync` method inside it. I prefer a domain service method for creating entities to keep the construction logic close to the other domain logic related to the entity.

Do not save the new entity to the database inside the `Factory` method and leave it to the client code (generally, an application service method). The `Factory` method's responsibility is to create the object and no more (think of it as an advanced constructor – an entity constructor cannot save the entity to the database, right?). The client code may need to perform additional operations on the entity before saving it. We will return to this topic in the next chapter.

Do not overuse factory methods and keep using simple public constructors wherever possible. Creating a valid entity is important, but it is just the beginning of an entity's lifecycle. In the next section, we will see how to change an entity's state in a controlled way.

Implementing business logic and constraints

An entity is responsible for keeping itself always valid. In addition to the constructor that ensures the entity is valid and consistent when it's first created, we can define methods on the entity class to change its properties in a controlled way.

As a simple rule, if changing a property's value has pre-conditions, we should make its setter private and provide a method to change its value by implementing the necessary business logic and validating the provided value.

See the Description property of the Event class:

```
public class Event : FullAuditedAggregateRoot<Guid>
{
    ...

    public string Description { get; private set; }

    public Event SetDescription(string description)
    {
        Description = Check.NotNullOrWhiteSpace(
            description, nameof(description),
            EventConsts.MaxDescriptionLength,
            EventConsts.MinDescriptionLength);
        return this;
    }
}
```

The Description property's setter is private. We provide the SetDescription method as the only way to change its value. In this method, we validate the description value: it should be a string that has a length of more than 50 (MinDescriptionLength) and less than 2000 (MaxDescriptionLength). These constants are defined in the *EventHub.Domain.Shared* project, so we can reuse them in DTOs, as we will see in the next chapter.

> **Data Annotation Attributes on Entity Properties**
>
> You may ask whether we can use `[Required]` or `[StringLength]` attributes on the `Description` property instead of creating a `SetDescription` method and manually performing the validation. Such attributes require another system that performs the validation. For example, EF Core can validate the properties based on these data annotation attributes while saving the entity to the database. However, that's not enough because, in that way, the entity could be invalid until we try to save it to the database. The entity should always be valid!

Let's see a more complex example, again from the `Event` class:

```
public Event AddSession(Guid trackId, Guid sessionId,
    string title, DateTime startTime, DateTime endTime,
    string description, string language)
{
    if (startTime < this.StartTime || this.EndTime <
        endTime)
    {
        throw new BusinessException(EventHubErrorCodes
            .SessionTimeShouldBeInTheEventTime);
    }

    var track = GetTrack(trackId);
    track.AddSession(sessionId, title, startTime, endTime,
                     description, language);
    return this;
}
private Track GetTrack(Guid trackId)
{
    return Tracks.FirstOrDefault(t => t.Id == trackId) ??
        throw new EntityNotFoundException(typeof(Track),
                                          trackId);
}
```

The `AddSession` method accepts a `trackId` parameter since a session should belong to a track. It also accepts the `sessionId` of the new session (getting it as a parameter to let the client use the `IGuidGenerator` service to create the value). The remaining parameters are the required properties of the new session.

The AddSession method first checks whether the new session is within the event's time range, then finds the right track (throws an exception otherwise) and delegates the remaining work to the track object. Let's see the track.AddSession method:

```
internal Track AddSession(Guid sessionId, string title,
    DateTime startTime, DateTime endTime,
    string description, string language)
{
    if (startTime > endTime)
    {
        throw new BusinessException(EventHubErrorCodes
            .EndTimeCantBeEarlierThanStartTime);
    }
    foreach (var session in Sessions)
    {
        if (startTime.IsBetween(session.StartTime,
            session.EndTime) ||
            endTime.IsBetween(session.StartTime,
            session.EndTime))
        {
            throw new BusinessException(EventHubErrorCodes
                .SessionTimeConflictsWithAnExistingSession);
        }
    }
    Sessions.Add(new Session(sessionId, Id, title,
                startTime, endTime, description));
    return this;
}
```

First of all, this method is internal to prevent using it out of the domain layer. It is always used by the Event.AddSession method shown earlier in this section.

The Track.AddSession method loops through all the current sessions to check whether any session time conflicts with the new session. If there's no problem, it adds the session to the track.

Returning this (the event object) from a setter method is a good practice since it allows us to chain the setters, for example, eventObject.SetTime(...).
SetDescription(...).

Both of the example methods used the properties on the event object and did not depend on any external object. What if we need to use an external service or repository to implement the business rule?

Using external services in entity methods

Sometimes, the business rule you want to apply needs to use external services. Entities cannot inject service dependencies because of technical and design restrictions. If you need to use a service in an entity method, the proper way to do this is to get that service as a parameter.

Assume that we have a business rule for the event capacity: you cannot decrease the capacity to lower than the currently registered user count. A null capacity value means there is no registration limitation.

See the following implementation on the Event class:

```
public async Task SetCapacityAsync(
    IRepository<EventRegistration, Guid>
        registrationRepository, int? capacity)
{
    if (capacity.HasValue)
    {
        var registeredUserCount = await
            registrationRepository.CountAsync(x =>
                x.EventId == @event.Id);
        if (capacity.Value < registeredUserCount)
        {
            throw new BusinessException(
            EventHubErrorCodes
            .CapacityCanNotBeLowerThanRegisteredUserCount);
        }
    }

    this.Capacity = capacity;
}
```

The `SetCapacityAsync` method uses a repository object to execute a database query to get the currently registered user count. If the count is higher than the new capacity value, then it throws an exception. The `SetCapacityAsync` method is async since it performs an async database call. The client (generally an application service method) is responsible for injecting and passing the repository service to this method.

The `SetCapacityAsync` method guarantees implementing the business rule because the `Capacity` property's setter is `private`, and this method is the only way to change it.

You can get external services into methods as parameters, as shown in this example. However, that approach makes the entity dependent on external services, making it complicated and harder to test. It also violates the single responsibility principle and mixes the business logic of different aggregates (`EventRegistration` is another aggregate root).

There is a better way to implement the business logic that depends on external services or works on multiple aggregates: domain services.

Implementing domain services

A domain service is another class in which we implement the domain rules and constraints. Domain services are typically needed when we need to work with multiple aggregates, and the business logic doesn't properly fit into any of these aggregates. Domain services are also used when we need to consume other services and repositories since they can use the dependency injection system.

Let's re-implement the `SetCapacityAsync` method (in the previous section) as a domain service method:

```
public class EventManager : DomainService
{
    ...
    public async Task SetCapacityAsync(Event @event,
                                       int? capacity)
    {
        if (capacity.HasValue)
        {
            var registeredUserCount = await
                _eventRegistrationRepository.CountAsync(
                    x => x.EventId == @event.Id);
```

```
            if (capacity.Value < registeredUserCount)
            {
                throw new BusinessException(
                EventHubErrorCodes.CapacityCanNotBeLower
                ThanRegisteredUserCount);
            }
        }
        @event.Capacity = capacity;
    }
}
```

In this case, we've injected `IRepository<EventRegistration, Guid>` into the `EventManager` domain service (see the source code for all the details) and got the `Event` object as a parameter. The setter of the `Event.Capacity` property is now `internal` so that it can be set only in the domain layer, in the `EventManager` class.

A domain service method should be fine-grained: it should make a small (yet meaningful and consistent) change to the aggregate. The application layer then combines these small changes to perform different use cases.

We will explore application services in the next chapter. However, I find it useful to show an example application service method here that updates multiple properties on an event in a single request:

```
public async Task UpdateAsync(Guid id,
                                UpdateEventDto input)
{
    var @event = await _eventRepository.GetAsync(id);
    
    @event.SetTitle(input.Title);
    @event.SetTime(input.StartTime, input.EndTime);
    await _eventManager.SetCapacityAsync(@event,
                                        input.Capacity);
    @event.Language = input.Language;
    
    await _eventRepository.UpdateAsync(@event);
}
```

The UpdateAsync method takes a DTO that contains the properties to be updated. It first retrieves the Event object from the database as a single unit, then uses SetTitle and SetTime methods on the Event object. These methods internally validate the provided values and properly change the property values.

The UpdateAsync method then uses the domain service method, eventManager.SetCapacity, to change the capacity value.

We directly set the Language property since it has a public setter and no business rule (it even accepts the null value). Do not create setter methods if they have no business rules or constraints. Also, do not create domain service methods simply to change the entity properties without any business logic.

The UpdateAsync method finally uses the repository to update the Event entity in the database.

Domain Service Interfaces

You do not need to introduce interfaces (such as IEventManager) for domain services since they are essential parts of the domain and should not be abstracted. However, if you want to mock domain services in unit tests, you may still want to create interfaces.

A domain service method should not update the entity as a common principle. In this example, we set the Language property after calling the SetCapacityAsync method. If SetCapacityAsync updates the entity, we end up with two database update operations, which would be inefficient.

As another good practice, accept the entity object as the parameter (as we've done in the SetCapacityAsync method) rather than its id value. If you accept its id value, you need to retrieve the entity from the database inside the domain service. This approach makes the application code load the same entity multiple times in different places in the same request (use case), which is inefficient and leads to bugs. Leave that responsibility to the application layer.

A specific type of domain service method is the factory method to create aggregates, explained in the *Using services to create aggregates* section. Declare factory methods only if a public constructor on the aggregate root cannot implement the business constraints. This may be the case if checking the business constraint requires the use of external services.

We've used repositories in many places so far. The next section explains the implementation details of repositories.

Implementing repositories

To remember the definition, a repository is a collection-like interface used to access the domain objects stored in the data persistence system. It hides the complexity of data access logic behind a simple abstraction.

There are some main rules for implementing repositories:

- Repository interfaces are defined in the domain layer, so the domain and application layers can use them. They are implemented in the infrastructure (or database provider integration) layer.

- Repositories are created for aggregate root entities but not for sub-collection entities. That is because the sub-collection entities should be accessed over the aggregate root. Typically, you have a repository for each aggregate root.

- Repositories work with domain objects, not DTOs.

- In an ideal design, repository interfaces should be independent of the database provider. So, do not get or return EF Core objects, such as `DbContext` or `DbSet`.

- Do not implement business logic inside repository classes.

ABP provides an implementation of the repository pattern out of the box. We explored how to use generic repositories and implement custom repositories in *Chapter 6, Working with the Data Access Infrastructure*. Here, I will discuss a few best practices.

The last rule in the preceding list, "Do not implement business logic inside repository classes," is the most important rule because the others are clear to understand. Implementing business logic inside a repository is generally a result of incorrectly considering business logic.

See the following example repository interface:

```
public interface IEventRepository : IRepository<Event,
                                                Guid>
{
    Task UpdateSessionTimeAsync(
        Guid sessionId, DateTime startTime, DateTime
            endTime);
    Task<List<Event>> GetNearbyEventsAsync();
}
```

At first sight, there is no problem; these methods are just performing some database operations. However, the devil is in the details.

The first method, `UpdateSessionTimeAsync`, changes the timing of a session in an event. If you remember, we had a business rule: a session's timing cannot overlap with another session on the same track. It also cannot overflow the event time range. If we implement that rule in the repository method, we duplicate that business validation because it was already implemented inside the `Event` aggregate. If we don't implement that validation, it is obviously a bug. In a true implementation, this logic should be done in the aggregate. The repository should only query and update the aggregate as a single unit.

The second method, `GetNearbyEventsAsync`, gets a list of events in the same city with the current user. The problem with this method is the *current user* is an application layer concept and requires an active user session. Repositories should not work with the current user. What if we want to reuse the same *nearby* logic in a background service where we don't have the current user in the current context? It is better to pass the city, date range, and other parameters to the method, so it simply brings the events. Entity properties are just values for the repositories. Repositories should not have any domain knowledge and should not use application layer features.

Repositories are fundamentally used to create, update, delete, and query entities. ABP's generic repository implementation provides most of the common operations out of the box. It also provides an `IQueryable` object that you can use to build and execute queries using LINQ. However, building complex queries at the application layer mixes your application logic with the data querying logic that should ideally be at the infrastructure layer.

See the following example method, which uses `IRepository<Event, Guid>` to get a list of events that a given user has spoken at:

```
public async Task<List<Event>> GetSpokenEventsAsync(Guid
                                                    userId)
{
    var queryable =
        await _eventRepository.GetQueryableAsync();
    var query = queryable.Where(x => x.Tracks
        .Any(track => track.Sessions
        .Any(session => session.Speakers
        .Any(speaker => speaker.UserId == userId))));
    return await AsyncExecuter.ToListAsync(query);
}
```

In the first line, we are obtaining an `IQueryable<Event>` object. Then we are using the `Where` method to filter the events. Finally, we are executing the query to get the event list.

The problem with writing such queries into application services is leaking querying logic to the application layer and making it impossible to reuse the querying logic when we need it somewhere else. To overcome the problem, we generally create a custom repository method to query the events:

```
public interface IEventRepository : IRepository<Event,
                                                Guid>
{
    Task<List<Event>> GetSpokenEventsAsync(Guid userId);
}
```

Now, we can use this custom repository method anywhere we need to get the events at which a user was a speaker.

Creating custom repository methods is a good approach. But, with this approach, we have a lot of similar methods once the application grows. Assume that we wanted to get the event list in a specified date range, and we've added one more method:

```
public interface IEventRepository : IRepository<Event,
                                                Guid>
{
    Task<List<Event>> GetSpokenEventsAsync(Guid userId);
    Task<List<Event>> GetEventsByDateRangeAsync(DateTime
        minDate, DateTime maxDate);
}
```

What if we want to query the events with a date range and speaker filter? Create another method as shown in the following code block:

```
Task<List<Event>> GetSpokenEventsByDateRangeAsync(Guid userId,
DateTime minDate, DateTime maxDate)
```

Actually, ABP provides the `GetListAsync` method, which takes an expression. So, we could remove all these methods and use the `GetListAsync` method with an arbitrary predicate.

The following example uses the `GetListAsync` method to get a list of events at which a user is a speaker in the next 30 days:

```
public async Task<List<Event>> GetSpokenEventsAsync(Guid
                                                    userId)
{
    var startTime = Clock.Now;
    var endTime = Clock.Now.AddDays(30);
    return await _eventRepository.GetListAsync(x =>
        x.Tracks
        .Any(track => track.Sessions
            .Any(session => session.Speakers
                .Any(speaker => speaker.UserId == userId)))
        && x.StartTime > startTime && x.StartTime <=
            endTime
    );
}
```

However, we've returned to the previous problem: mixing the querying complexity with the application code. Also, isn't the query getting hard to understand? You know, in real life, we have queries with much more complexity.

Completely getting rid of complex queries may not be possible, but the next section offers an interesting solution to that problem: the specification pattern!

Building specifications

A specification is a named, reusable, combinable, and testable class to filter domain objects based on business rules. In practice, we can easily encapsulate filter expressions as reusable objects.

In this section, we will begin with the most simple, parameterless specifications. We will then see more complex, parameterized specifications. Finally, we will learn how to combine multiple specifications to create a more complex specification.

Parameterless specifications

Let's begin with a very simple specification class:

```
public class OnlineEventSpecification :
    Specification<Event>
{
    public override Expression<Func<Event, bool>>
        ToExpression()
    {
        return x => x.IsOnline == true;
    }
}
```

`OnlineEventSpecification` is used to filter online events, which means it selects an event if it is an online event. It is derived from the base `Specification<T>` class provided by ABP Framework to create specification classes easily. We override the `ToExpression` method to filter the event objects. This method should return a lambda expression that returns `true` if the given `Event` entity (here, the x object) satisfies the condition (we could simply write `return x => x.IsOnline`).

Now, if we want to get a list of online events, we can just use the repository's `GetListAsync` method with a specification object:

```
var events = _eventRepository
    .GetListAsync(new OnlineEventSpecification());
```

Specifications are implicitly converted to expressions (remember, the `GetListAsync` method can get an expression). If you want to explicitly convert them, you can call the `ToExpression` method:

```
var events = _eventRepository
    .GetListAsync(
        new OnlineEventSpecification().ToExpression());
```

So, we can use a specification wherever we can use an expression. In this way, we can encapsulate expressions as named, reusable objects.

The `Specification` class provides another method, `IsSatisfiedBy`, to test a single object. If you have an `Event` object, you can easily check whether it is an online event or not:

```
Event evnt = GetEvent();
if (new OnlineEventSpecification().IsSatisfiedBy(evnt))
{
    // ...
}
```

In this example, we have somehow obtained an `Event` object, and we want to check whether it is online. `IsSatisfiedBy` gets an `Event` object and returns `true` if that object satisfies the condition. I accept that this example seems absurd because we could simply write `if(evnt.IsOnline)`. Such a simple specification was not necessary. However, in the next section, we will see more complex examples to make it much clearer.

Parameterized specifications

Specifications can have parameters to be used in filtering expressions. See the following example:

```
public class SpeakerSpecification : Specification<Event>
{
    public Guid UserId { get; }

    public SpeakerSpecification(Guid userId)
    {
        UserId = userId;
    }

    public override Expression<Func<Event, bool>>
        ToExpression()
    {
        return x => x.Tracks
            .Any(t => t.Sessions
                .Any(s => s.Speakers
                    .Any(sp => sp.UserId == UserId)));
```

```
        }
}
```

We've created a parameterized specification class that checks whether the given user is a speaker at an event. Once we have that specification class, we can filter the events as shown in the following code block:

```
public async Task<List<Event>> GetSpokenEventsAsync(Guid
                                                    userId)
{
    return await _eventRepository.GetListAsync(
        new SpeakerSpecification(userId));
}
```

Here, we've just reused the GetListAsync method of the repository by providing a new SpeakerSpecification object. From now on, we can reuse this specification class if we need the same expression later, in another place in our application, without needing to copy/paste the expression. If we need to change the condition later, all those places will use the updated expression.

If we need to check whether a user is a speaker at the given Event, we can reuse the SpeakerSpecification class by calling its IsSatisfiedBy method:

```
Event evnt = GetEvent();
if (new SpeakerSpecification(userId).IsSatisfiedBy(evnt))
{
    // ...
}
```

Specifications are powerful to create named and reusable filters, but they have another power too: combining specifications to create a composite specification object.

Combining specifications

It is possible to combine multiple specifications using operator-like `And`, `Or`, and `AndNot` methods, or to reverse a specification with the `Not` method.

Assume that I want to find the events where a given user is a speaker, and the event is online:

```
var events = _eventRepository.GetListAsync(
    new SpeakerSpecification(userId)
        .And(new OnlineEventSpecification())
        .ToExpression()
);
```

In this example, I combined `SpeakerSpecification` and `OnlineEventSpecification` objects to create a composite specification object. Explicitly calling the `ToExpression` class is necessary in this case because C# doesn't support implicitly converting from interfaces (the `And` method returns an `ISpecification<T>` reference).

The following example finds the in-person (offline) events in the next 30 days where the given user is a speaker:

```
var events = _eventRepository.GetListAsync(
    new SpeakerSpecification(userId)
        .And(new DateRangeSpecification(Clock.Now,
            Clock.Now.AddDays(30)))
        .AndNot(new OnlineEventSpecification())
        .ToExpression()
);
```

In this example, we've reversed the `OnlineEventSpecification` object's filtering logic with the `AndNot` method. We've also used a `DateRangeSpecification` object that we haven't defined yet. It is a good exercise for you to implement yourself.

An interesting example could be extending the `AndSpecification` class to create a specification class that combines two specifications:

```
public class OnlineSpeakerSpecification :
    AndSpecification<Event>
{
    public OnlineSpeakerSpecification(Guid userId)
```

```
        : base(new SpeakerSpecification(userId),
            new OnlineEventSpecification())
    {
    }
}
```

The `OnlineSpeakerSpecification` class in this example combines the `SpeakerSpecification` class and the `OnlineEventSpecification` class, and can be used whenever you want to use a specification object.

> **When to Use Specifications**
>
> Specifications are especially useful if they filter objects based on domain rules that can be changed in the future, so you don't want to duplicate them everywhere. You do not need to define specifications for the expressions you are using just for reporting purposes.

The next section explains how to use domain events to publish notifications.

Publishing domain events

Domain events are used to inform other components and services about an important change to a domain object so that they can take action.

ABP Framework provides two types of event buses to publish domain events, each with a different purpose:

- The **local event bus** is used to notify the handlers in the same process.
- The **distributed event bus** is used to notify the handlers in the same or different processes.

Publishing and handling events are pretty easy with ABP Framework. The next section shows how to work with the local event bus, and then we will look at the distributed event bus.

Using the local event bus

A local event handler is executed in the same unit of work (in the same local database transaction). If you are building a monolith application or want to handle events in the same service, the local event bus is fast and safe to use because it works in the same process.

Assume that you want to publish a local event when an event's time changes, and you have an event handler that sends emails to the registered users about the change.

See the simplified implementation of the SetTime method of the Event class:

```
public void SetTime(DateTime startTime, DateTime endTime)
{
    if (startTime > endTime)
    {
        throw new BusinessException(EventHubErrorCodes
            .EndTimeCantBeEarlierThanStartTime);
    }
    StartTime = startTime;
    EndTime = endTime;
    if (!IsDraft)
    {
        AddLocalEvent(new EventTimeChangedEventData(this));
    }
}
```

In this example, we are adding a local event, which will be published while updating the entity. ABP Framework overrides EF Core's SaveChangesAsync method to publish the events (for MongoDB, it is done in the repository's UpdateAsync method).

Here, EventTimeChangedEventData is a plain class that holds the event data:

```
public class EventTimeChangedEventData
{
    public Event Event { get; }
    public EventTimeChangedEventData(Event @event)
    {
        Event = @event;
    }
}
```

Published events can be handled by creating a class that implements the ILocalEventHandler<TEventData> interface:

```
public class UserEmailingHandler :
    ILocalEventHandler<EventTimeChangedEventData>,
```

```
    ITransientDependency
{
    public async Task HandleEventAsync(
        EventTimeChangedEventData eventData)
    {
        var @event = eventData.Event;
        // TODO: Send email to the registered users!
    }
}
```

The `UserEmailingHandler` class can inject any service (or repository) to get a list of the registered users, and then send an email to inform them about the time change. You may have multiple handlers for the same event. If any handler throws an exception, the main database transaction is rolled back since the event handler is executed in the same database transaction.

Events can be published in entities, as shown in the previous examples. They can also be published using the `ILocalEventBus` service.

Let's assume that we don't publish the `EventTimeChangedEventData` event inside the `Event` class but want to publish it in an arbitrary class that can utilize the dependency injection system. See the following example application service:

```
public class EventAppService : EventHubAppService,
    IEventAppService
{
    private readonly IRepository<Event, Guid>
        _eventRepository;
    private readonly ILocalEventBus _localEventBus;

    public EventAppService(
        IRepository<Event, Guid> eventRepository,
        ILocalEventBus localEventBus)
    {
        _eventRepository = eventRepository;
        _localEventBus = localEventBus;
    }

    public async Task SetTimeAsync(
```

```
            Guid eventId, DateTime startTime, DateTime endTime)
    {
        var @event =
            await _eventRepository.GetAsync(eventId);
        @event.SetTime(startTime, endTime);
        await _eventRepository.UpdateAsync(@event);
        await _localEventBus.PublishAsync(
            new EventTimeChangedEventData(@event));
    }
}
```

The `EventAppService` class injects the repository and `ILocalEventBus` service. In the `SetTimeAsync` method, we are using the local event bus to publish the same event.

The `PublishAsync` method of the `ILocalEventBus` service immediately executes the event handlers. If any event handler throws an exception, you directly get the exception since the `PublishAsync` method doesn't handle exceptions. So, if you don't catch the exception, the whole unit of work is rolled back.

It is better to publish the events in the entities or the domain services. If we publish the `EventTimeChangedEventData` event in the `Event` class's `SetTime` method, it is guaranteed to publish the event in any case. However, if we publish it in an application service, like in the last example, we may forget to publish the event in another place that changes the event times. Even if we don't forget, we will have duplicated code that is harder to maintain and open to potential bugs.

Local events are especially useful to implement side-effects such as taking extra action when the state of an object changes. It is perfect for de-coupling and integrating different aspects of the system. In this section, we have used it to send a notification email to the registered users when an event's time changes. However, it should not be misused by distributing business logic flow into event handlers and making the whole process hard to follow.

In the next section, we will see the second type of event bus.

Using the distributed event bus

In the distributed event bus, an event is published through a message broker service, such as RabbitMQ or Kafka. If you are building a microservice/distributed solution, the distributed event bus can asynchronously notify the handlers in other services.

Using the distributed event bus is pretty similar to the local event bus, but it is important to understand the differences and limitations.

Let's assume that we want to publish a distributed event in the `Event` class's `SetTime` method when the event's time changes:

```
public void SetTime(DateTime startTime, DateTime endTime)
{
    if (startTime > endTime)
    {
        throw new BusinessException(EventHubErrorCodes
            .EndTimeCantBeEarlierThanStartTime);
    }
    StartTime = startTime;
    EndTime = endTime;
    if (!IsDraft)
    {
        AddDistributedEvent(new EventTimeChangedEto
        {
            EventId = Id, Title = Title,
            StartTime = StartTime, EndTime = EndTime
        });
    }
}
```

Here, we call the `AddDistributedEvent` method to publish the event (it is published when the entity is updated in the database). As an important difference from the local event, we are not passing the entity (`this`) object as the event data but copying some properties to a new object. That new object will be transferred between processes. It will be serialized in the current process and deserialized in the target process (ABP Framework handles the serialization and deserialization for you). So, creating a DTO-like object that carries only the required properties rather than the full object is better. The **Eto** (**Event Transfer Object**) suffix is the naming convention that we suggest but it is not necessary to use it.

The `AddDistributedEvent` (and `AddLocalEvent`) method is only available in the aggregate root entities, not for sub-collection entities. However, publishing a distributed event in an arbitrary service is still possible using the `IDistributedEventBus` service:

```
await _distributedEventBus.PublishAsync(
    new EventTimeChangedEto
    {
        EventId = @event.Id,
        Title = @event.Title,
        StartTime = @event.StartTime,
        EndTime = @event.EndTime
    });
```

Inject the `IDistributedEventBus` service and use the `PublishAsync` method – that's all.

The application/service that wants to get notified can create a class that implements the `IDistributedEventHandler<T>` interface, as shown in the following code block:

```
public class UserEmailingHandler :
    IDistributedEventHandler<EventTimeChangedEto>,
    ITransientDependency
{
    public Task HandleEventAsync(EventTimeChangedEto
                                    eventData)
    {
        var eventId = eventData.EventId;
        // TODO: Send email to the registered users!
    }
}
```

The event handler can use all the properties of the `EventTimeChangedEto` class. If it needs more data, you can add it to the `ETO` class. Alternatively, you can query the details from the database or perform an API call to the corresponding service in a distributed scenario.

Summary

This chapter covered the first part of implementing DDD. We've explored the domain layer building blocks and understood their design and implementation practices using ABP Framework.

The aggregate is the most fundamental DDD building block, and the way we change an aggregate's state is very important and needs care. An aggregate should preserve its validity and consistency by implementing business rules. It is essential to draw aggregate boundaries correctly.

On the other hand, domain services are useful for implementing the domain logic that touches multiple aggregates or external services. They work with the domain objects, not DTOs.

The repository pattern abstracts the data access logic and provides an easy-to-use interface to other services in the domain and application layers. It is important not to leak your business logic into repositories. The specification pattern is a way to encapsulate data filtering logic. You can reuse and combine them when you want to select business objects.

Finally, we explored how we can publish and subscribe to domain events with ABP Framework. Domain events are used to react to changes to domain objects in a loosely coupled way.

The next chapter will continue with the building blocks, this time in the application layer. It will also discuss the differences between the domain and application layers with practical examples.

11

DDD – The Application Layer

The previous chapter explained the domain-layer building blocks with details. The domain layer is used to implement the core, application-independent domain logic of the solution. However, we also need some applications to interact with that domain logic, such as a web or mobile application. The application layer is responsible for implementing the business logic of such applications without depending on the **user interface** (**UI**) technology used in the presentation layer. We keep the domain layer isolated from the presentation technology by encapsulating it with the application services.

In this chapter, we will learn how to design and implement the application services and **data transfer objects** (**DTOs**) with ABP Framework. We will also understand the differences between domain-layer and application-layer responsibilities.

This chapter covers the following topics:

- Implementing application services
- Designing DTOs
- Understanding the responsibilities of the layers

Technical requirements

You can clone or download the source code of the *EventHub* project from GitHub: `https://github.com/volosoft/eventhub`.

If you want to run the solution in your local development environment, you need an **integrated development environment** (**IDE**)/editor (such as Visual Studio) to build and run ASP.NET Core solutions. Also, if you want to create ABP solutions, you need to have the ABP **command-line interface** (**CLI**) installed, as explained in *Chapter 2, Getting Started with ABP Framework*.

Implementing application services

An application service is a stateless class used by the presentation layer to perform use cases of the application. It orchestrates the domain objects to achieve the business operation. Application services get and return DTOs instead of entities.

An application service method is considered a work unit (meaning all database operations—all succeed or all fail as a group, as covered in *Chapter 6, Working with the Data Access Infrastructure*), which ABP Framework automatically does. A typical flow of an application service method includes the following steps:

1. Get the necessary aggregates from the repositories using the input parameters and the current context.

2. Implement the use case by coordinating the aggregates, domain services, and other domain objects, and delegating the work to them.

3. Update the changed aggregates in the database using the repositories.

4. Optionally, return a resulting DTO to the client (typically, to the presentation layer).

> **About Updating Changed Objects**
>
> In fact, *Step 3* is not necessary if you use **Entity Framework Core** (**EF Core**), since EF Core has a change-tracking system that can automatically determine changed objects and update them in the database at the end of the **unit of work** (**UoW**). So, if you have no problem relying on EF Core features, you can skip *Step 3*.

Let's see the `AddSessionAsync` application service method in the following example:

```
public class EventAppService : ApplicationService,
    IEventAppService
{
```

```
...
[Authorize]
public async Task AddSessionAsync(Guid id,
                                        AddSessionDto input)
{
    var @event = await _eventRepository.GetAsync(id);
    @event.AddSession(
        input.TrackId, GuidGenerator.Create(),
        input.Title,
        input.StartTime, input.EndTime,
        input.Description, input.Language
    );
    await _eventRepository.UpdateAsync(@event);
}
}
```

This method is a simple application service method. It is used to add a new session to an event. It first gets the related `Event` aggregate from the database. Then, it uses the `AddSession` method of the `Event` class to delegate the actual business operation to the domain layer. It finally updates the changed `Event` object in the database. We will see the `AddSessionDto` class in the *Designing DTOs* section of this chapter.

Let's see a more complex example that creates a new event, as follows:

```
[Authorize]
public async Task<EventDto> CreateAsync(CreateEventDto
                                                    input)
{
    var organization = await _organizationRepository
        .GetAsync(input.OrganizationId);

    if (organization.OwnerUserId != CurrentUser.GetId())
    {
        throw new AbpAuthorizationException(
        L["EventHub:
            NotAuthorizedToCreateEventInThisOrganization",
            organization.DisplayName]
        );
```

```
    }

    var @event = await _eventManager.CreateAsync(
        organization, input.Title,
        input.StartTime, input.EndTime, input.Description);

    await _eventManager.SetLocationAsync(@event,
        input.IsOnline, input.OnlineLink, input.CountryId,
        input.City);
    await _eventManager.SetCapacityAsync(@event,
                                        input.Capacity);
    @event.Language = input.Language;

    if (input.CoverImageContent != null &&
        input.CoverImageContent.Length > 0)
    {
        await SaveCoverImageAsync(
            @event.Id, input.CoverImageContent);
    }

    await _eventRepository.InsertAsync(@event);
    return ObjectMapper.Map<Event, EventDto>(@event);
}
```

The `CreateAsync` method gets a `CreateEventDto` object from the UI layer that carries the new event data and is a good example of creating new entities. Let's investigate how it was implemented.

It first gets the `organization` object from the database and compares its owner's **identifier** (**ID**) with the current user's ID. It throws an `AbpAuthorizationException` exception if the user doesn't meet the condition. Authorization is an application-layer responsibility, and the authorization rules can be different in different applications. For example, in the admin application, an admin user can create events on behalf of any user without checking the organization's ownership.

The `CreateAsync` method then uses the `eventManager` domain service to create a new `Event` object with the minimum required properties.

We've created an `Event` object, but our work hasn't been completed yet. `CreateEventDto` has some optional properties that the user may set. We are again using the `eventManager` domain service to set the location of the event. We are then directly setting the `Language` property of `Event` because there is no business rule to set it; it has a public setter, and the value can even be `null`.

The `CreateAsync` method continues using the `eventManager` class to set the event capacity by checking the core domain rules. It also saves the event's cover image if it was provided.

Until that point, the `Event` object hasn't been saved to the database. All the operations are performed on an in-memory object. The domain service doesn't save changes to the database since it is the application layer's responsibility to do this. If the domain service methods saved their changes, we would end up with one insert and three update operations in the database. With the current implementation, the `CreateAsync` method uses the `InsertAsync` method of the repository and saves the object at the end of the method with a single database operation.

As you see in the example application service definitions, application service methods use DTO classes to get data from the upper layer (typically, the presentation layer) and return data to the upper layer. The next section introduces DTO design considerations and best practices.

Designing DTOs

A DTO is a simple object used to transfer data between the presentation and the application layers. Let's start by seeing the basic principles of designing DTO classes.

Designing DTO classes

There are some fundamental principles to follow while defining DTO classes, as outlined here:

- DTOs should not contain any business logic; they are just for data transfer.

- DTO objects should be serializable because most of the time, they are transferred over the wire. Typically, they have a parameterless constructor, and all of their properties have public getters and setters.

- DTO classes should not inherit from the entities or use entity types as their properties.

The following DTO class is used to store the data while adding a new session to an existing track of an event:

```
public class AddSessionDto
{
    [Required]
    [StringLength(SessionConsts.MaxTitleLength,
        MinimumLength = SessionConsts.MinTitleLength)]
    public string Title { get; set; }
    [Required]
    [StringLength(SessionConsts.MaxDescriptionLength,
        MinimumLength =
            SessionConsts.MinDescriptionLength)]
    public string Description { get; set; }
    public Guid TrackId { get; set; }
    public DateTime StartTime { get; set; }
    public DateTime EndTime { get; set; }
    public string Language { get; set; }
}
```

The `AddSessionDto` class has no method, so it has no business logic. All its properties have public getters and setters. The `AddSessionDto` class does not define any constructor, so it has an implicit public parameterless constructor.

The `Title` and `Description` properties of the `AddSessionDto` class have validation attributes, such as `Required` and `StringLength`.

The next section discusses validating input DTOs.

Validating input DTOs

There are a few ways to validate a DTO object when it is used as a parameter to an application service method, as outlined here:

- We can use data annotation attributes, such as `Required`, `StringLength`, and `Range`.

- We can implement an `IValidatableObject` interface for the DTO class and perform additional validation logic in the `Validate` method.

- We can use third-party libraries to validate a DTO object. For example, ABP integrates to the `FluentValidation` library to separate the validation logic from the DTO class and perform advanced validation logic.

Whichever approach you follow (you can use all together for a DTO class), ABP automatically checks these validation rules and throws a validation exception in case of an invalid value. So, your application service method is always executed with a valid DTO object. See the *Validating user inputs* section of *Chapter 7, Exploring Cross-Cutting Concerns,* for all the details of ABP's validation infrastructure.

The validation logic on the DTO class (or in the `FluentValidation` validator class) should only be a formal validation. That means you can check if the given input is supplied and well formatted. However, it should not contain a domain validation. For example, do not try to check if the given start and end dates conflict with another session on the same track. Such validation logic should be implemented in the domain layer, typically in the entity or a domain service class.

Another common task with DTOs is mapping them to other objects.

Object-to-object mapping

We use entities inside the domain and application layers and use DTOs to communicate with the upper layers. This approach leads us to create DTO classes similar to entity classes and convert entity objects to DTO objects. If the entity class has a few properties, then creating a corresponding DTO object can be manually done by copying properties one by one. However, entity classes grow over time, and writing and maintaining a manual mapping code becomes tedious and error-prone.

ABP Framework provides an `IObjectMapper` service that is used to convert similar objects to each other. See the following application service method:

```
public async Task<EventDto> GetAsync(Guid id)
{
    Event eventEntity =
        await _eventRepository.GetAsync(id);
    return ObjectMapper.Map<Event, EventDto>(eventEntity);
}
```

This method simply returns an `EventDto` object by converting it from the `Event` object using the `IObjectMapper` service. `EventDto` has a lot of properties, and manually creating it would result in a long code block. `IObjectMapper` is an abstraction and is implemented using the `AutoMapper` library when you create a new ABP solution. If you want to use the preceding code, you should first define the `AutoMapper` mapping configuration.

> **Object-to-Object Mapping Documentation**
>
> The topic of object-to-object mapping is not included in this book. However, we did use it in *Chapter 3, Step-By-Step Application Development,* while creating an example application. Please refer to ABP's documentation to fully understand the object-to-object mapping system: `https://docs.abp.io/en/abp/latest/Object-To-Object-Mapping`.

While using the object mapper is pretty simple, we should use it carefully. Object mapping libraries mostly rely on naming conventions. They automatically map same-named properties, while we can configure the mappings manually.

One possible problem may occur when you refactor entities but do not update the corresponding DTOs or the mapping code. The `AutoMapper` library has a concept named configuration validation. It validates the mapping configuration on the application startup and throws an exception if it detects a mapping configuration problem. I suggest enabling it for your application. See the `AutoMapper` documentation to learn about configuration validation: `https://docs.automapper.org/en/stable/Configuration-validation.html`.

Object-to-object mapping is really useful when you map your entities to DTOs. However, do not map input DTOs to entities. There are some technical and design reasons behind this suggestion, as outlined here:

- Do you remember the *Implementing entity constructors* section of the previous chapter? Entity classes typically have primary constructors to get the required properties and create a valid entity. Auto-mapping operations generally need an empty constructor on the target class, so the mapping fails.

- Some properties on the entities are designed with private setters. You should use entity methods to change these property values to apply some business rules. Directly copying their values from a DTO object may violate the business rules.

- You should carefully validate and process the user input instead of blindly mapping to the entities.

The `CreateAsync` method explained in the *Implementing application services* section was a good example of creating an entity using an input DTO. It doesn't map the DTO to the entity but uses a domain service to create a valid entity and set optional properties.

In the next section, we will discuss some design practices for DTOs.

DTO design best practices

Creating DTOs seems simple at first—they are simple, indeed. However, once the application grows, you will have many DTO classes, and it becomes important to understand how to organize these classes. In the next sections, I will provide a few suggestions about DTOs to make your code base more maintainable and bug-free.

Do not define unused properties in input DTOs

No one defines an unused property in an input DTO, right? Unfortunately, this isn't true. I have seen a lot of code bases with this problem.

A property in a DTO class that is not used in the application service method is a perfect way to confuse the developers using that application service method and lead to building a buggy code base.

A probable reason for an unused property in a DTO class is that it was used before, but the application service method was changed and the developer has forgotten to remove it. We should care about that and always remove unused properties. If you care about backward compatibility because you are not the one building the client application, then declare an [Obsolete] attribute on that property, document the breaking change, and try to preserve the old behavior if the value is provided.

Having unused properties can be unavoidable if you violate the rule in the next section.

Do not reuse input DTOs

When you have too many application service methods, you may think that using some DTOs for multiple application service methods is a good idea to reduce the number of DTOs. However, if you do that, some properties will be used in some methods but not used in others.

It is a good practice to define a specialized input DTO for each application service method. Sometimes, it seems practical to reuse the same DTO class for two methods since they are almost the same. However, the application service methods will change over time, and the requirements will be different. Inheriting a DTO from another DTO is another way of reusing DTOs, but the problem is the same.

Code duplication is a better practice than coupling use cases in many scenarios. See the following example:

```
public interface IEventAppService : IApplicationService
{
    Task<EventDto> GetAsync(Guid id);
    Task CreateAsync(EventDto input);
    Task UpdateEventTimeAsync(EventDto input);
}
```

In this example, the `GetAsync` method returns an `EventDto` object that stores almost all of the event properties. The `CreateAsync` method reuses the same `EventDto` class. In general, reusing an output DTO as an input DTO is not good since some `EventDto` properties (such as `Id`, `UrlCode`, `RegisteredUserCount`, and `CreationTime`) are not expected to be sent by the client application on event creation but calculated on the server side. Finally, reusing the same `EventDto` class in the `UpdateEventTimeAsync` method is much worse since this method only uses the `Id`, `StartTime`, and `EndTime` properties.

A true DTO design is shown in the following example:

```
public interface IEventAppService : IApplicationService
{
    Task<EventDto> GetAsync(Guid id);
    Task CreateAsync(EventCreationDto input);
    Task UpdateEventTimeAsync(EventTimeUpdateDto input);
}
```

We've defined a separate DTO class for the `CreateAsync` and `UpdateEventTimeAsync` methods. In this way, any change in one DTO won't affect the other methods.

The design suggestions in the last two sections were for input DTOs. The next section explains the case for output DTOs.

About output DTOs

In practice, output DTOs are different than input DTOs. The *unused property* problem (of input DTOs) does not exist for output DTOs. Let's try to understand why an unused property is a problem for an input DTO. Imagine that we are calling a method and setting a property on the input DTO. We expect that it is being processed by the method and change the behavior. We get confused if the method doesn't use the property, and we don't see any behavior difference, whatever we set for that property. If a method parameter (or property of the parameter) is there, it should work just as expected.

However, for output DTOs, that's not the case. An application service may return more properties than the client currently needs. Thus, I mean that the application service method fills all the properties of an output DTO; if a property does exist in the DTO, it should always be filled regardless of whether it is being used by the client or not.

This approach has an advantage—when the client later needs those properties, we don't need to change the service class, making it easier to extend the UI. This is especially useful if the client is not your application or you want to open your **application programming interface (API)** to third-party clients whose needs can be different.

Since an output DTO may contain some properties that are not used by the client yet, we can reduce the number of output DTOs and reuse some DTO classes in multiple cases.

See the following example application service definition:

```
public interface IEventAppService : IApplicationService
{
    Task<EventDto> CreateAsync(CreateEventDto input);
    Task<EventDto> GetAsync(Guid id);
    Task<List<EventDto>> GetListAsync(PagedResultRequestDto
                                        input);
    Task<EventDto> AddSessionAsync(Guid id,
                                    AddSessionDto input);
}
```

All the methods on this example use the same output DTO class to represent an event, `EventDto`, while they get different input DTO objects.

> **Performance Considerations**
>
> Returning fine-tuned and minimal-output DTO objects can be needed for performance requirements, especially when returning large result sets. In such cases, you can define different DTO classes with the properties only needed for the related use cases.

We've covered the fundamental building blocks for implementing DDD with ABP Framework. The next section demonstrates some examples to understand the roles and responsibilities of the layers.

Understanding the responsibilities of the layers

Separating your business logic into application and domain layers allows you to create multiple applications on the same domain, as explained in the *Dealing with multiple applications* section of *Chapter 9, Understanding Domain-Driven Design*. Large systems typically have multiple applications and isolating the core domain from application-specific logic is a key principle to not mix the logic of different applications. Creating a separate application layer for each application makes it possible to design our application service methods best suited to different application requirements.

To successfully separate the application and domain layers, we should have a good understanding of each layer's responsibilities. In the last three chapters, I have already mentioned these responsibilities while explaining the DDD building blocks. In the next sections, I will summarize these responsibilities to understand them better.

Authorizing users

Authorization is used to allow or prevent using a certain application functionality by a user (on a UI) or a client application (on a machine-to-machine communication).

ABP provides a declarative way of checking user permissions with the `[Authorize]` attribute and an imperative way with the `IAuthorizationService` service. You can use these features to restrict access to desired functionalities in your application.

Authorization is the responsibility of the application layer and the upper layers (such as the presentation layer) because it highly depends on the clients and the users of the application.

For example, in the *EventHub* project, a user in the public web application can edit only their own events. However, an admin user in the admin application can edit any event without the ownership checking if they have the required permission. On the other hand, a background service may change an event's state without any authorization rule. These applications use the same domain layer, hence the same domain rules, but implement different authorization rules. So, it is good not to include the authorization logic in the domain layer to make it reusable in different cases.

Controlling the transaction

The UoW system's responsibility is to create a transaction scope for a use case (typically, a web request) and ensure that all the changes done in that use case are committed together. The UoW system was covered in the *Understanding the UoW system* section of *Chapter 6, Working with the Data Access Infrastructure*.

The scope of a use case is an application service method. An application service method may work with multiple domain services and aggregates and may make changes on the aggregates. The only way to ensure that all changes are committed together is to control the UoW system at the application service level, in the application layer. So, a UoW is an application-layer concept.

Validating the user input

As stated in the *Validating user inputs* section of *Chapter 7, Exploring Cross-Cutting Concerns*, a typical application has three levels of validation, as follows:

- *Client-side validation* is used to pre-validate the user input before sending data to the server. This type of validation is the responsibility of the presentation layer.

- *Server-side validation* is performed by the server to prevent incomplete, badly formatted, or malicious requests. We generally use data annotation attributes and other functionalities to validate DTO objects. Such validation is the responsibility of the application layer.

- *Business validation* is also performed in the server, implements your business rules, and keeps your business data consistent. Business validation is mostly done in the domain layer to force the same business rules in every application.

ABP provides good infrastructure and gracefully integrates to ASP.NET Core services to easily perform formal validation logic. You can implement business validation rules and constraints in the aggregate constructors, methods, and domain services, as explained in *Chapter 10, DDD – The Domain Layer*.

Working with the current user

ABP's `ICurrentUser` service is used to get information about the current user. The current user logic requires a stateful session system that stores user information and makes it available in every web request.

For ASP.NET Core applications, ABP uses the current principle that is based on the authentication ticket. An authentication ticket is created when the user logs in to the application. It is saved in a cookie to read in subsequent requests. It can be stored in local storage for a **single-page application (SPA)** and sent to the server in the **HyperText Transfer Protocol (HTTP)** header for every request.

The session/current user is a concept that is typically implemented in the presentation layer. It is usable in the application layer since the application layer is designed to be used by the presentation layer, so it can assume that there is a *current user* in the current context. However, the domain layer should be designed as independent from any application, so it should not work with the current user. In some applications, such as a background service or an integration application, there may not be a user at all.

The following application service method, from the *EventHub* project, is used to join a given organization by the current user:

```
[Authorize]
public async Task JoinAsync(Guid organizationId)
{
    var organization = await _organizationRepository
        .GetAsync(organizationId);
    var user = await
        _userRepository.GetAsync(CurrentUser.GetId());
    await _organizationMembershipManager.JoinAsync(
        organization, user);
}
```

First of all, this method is authorized. So, it is guaranteed that the method is called by a user that has already logged in to the application and `CurrentUser.GetId()` returns a valid user ID.

We are not accepting the user's ID as a method parameter; otherwise, any authenticated user could make any user a member of any organization. But we want every user to be able to join the organization themselves.

We get the organization and user aggregates from the repositories and delegate the work to the domain service (`organizationMembershipManager`). The domain service, in this way, is independent of the current user concept and also more reusable: it can work with any user, not only with the current one.

Summary

In this chapter, you've learned how to properly implement the application services and design the DTOs. I've covered DTO design in detail, such as validating input DTOs and mapping entities to DTOs, and provided suggestions based on best practices and my experience.

We've learned that mixing the responsibilities of the layers makes layering meaningless. We've investigated some fundamental responsibilities to understand at which layer we should implement these responsibilities.

As we're at the end of this chapter, we've completed the third part of the book. The purpose of this part was to demonstrate how you can implement **domain-driven design (DDD)** building blocks with ABP Framework. I've provided rules, best practices, and suggestions to make your code base more maintainable when you follow them.

The next part of the book will explore the UI and API development with ABP Framework. In the next chapter, we will learn the architectural structure and the fundamental features of the ABP Framework MVC/Razor Pages UI.

Part 4:
User Interface and
API Development

This part explains how to create user interfaces with the MVC (Razor Pages) and Blazor UI options, as well as creating HTTP APIs for remote clients.

In this part, we include the following chapters:

- *Chapter 12, Working with MVC/Razor Pages*
- *Chapter 13, Working with the Blazor WebAssembly UI*
- *Chapter 14, Building HTTP APIs and Real-Time Services*

12
Working with MVC/ Razor Pages

ABP Framework was designed to be modular, layered, and UI framework-agnostic. It is perfect for server-client architecture, and in theory, it can work with any kind of UI technology. The server-side uses the standard authentication protocols and provides standard-compliant HTTP APIs. You can use your favorite SPA framework and consume server-side APIs easily. In this way, you can utilize the entire server-side infrastructure of ABP Framework.

However, ABP Framework helps with your UI development too. It provides systems so that you can build modular user interfaces, UI themes, layouts, navigation menus, and toolbars. It makes your development process easier while working with data tables, modals, and forms or authenticating and communicating with the server.

ABP Framework is well integrated with, and provides startup solution templates for, the following UI frameworks:

- ASP.NET Core MVC/Razor Pages

- Blazor

- Angular

In the fourth section of this book, I will cover working with the MVC/Razor Pages and Blazor UI options. In this chapter, you will learn how ABP Framework's MVC/Razor Page infrastructure is designed and how it can help you with your regular UI development cycle.

I call this UI type MVC/Razor Pages as it supports both the MVC and Razor Pages approaches. You can even use both in a single application. However, since Razor Pages (introduced with ASP.NET Core 2.0) is Microsoft's recommended approach for new applications, all pre-built ABP modules, samples, and documents use the Razor Pages approach.

This chapter covers the following topics:

- Understanding the theming system

- Using bundling and minification

- Working with menus

- Working with Bootstrap tag helpers

- Creating forms and implementing validation

- Working with modals

- Using the JavaScript API

- Consuming HTTP APIs

Technical requirements

If you want to follow along with the examples in this chapter, you will need to have an IDE/editor that supports ASP.NET Core development. We will use the ABP CLI at some points, so you will need to install the ABP CLI, as explained in *Chapter 2, Getting Started with ABP Framework*. Finally, you will need to install Node.js v14+ to be able to install NPM packages.

You can download the example application from this book's GitHub repository: `https://github.com/PacktPublishing/Mastering-ABP-Framework`. It contains some of the examples provided in this chapter.

Understanding the theming system

UI styling is the most customized part of an application, and you have plenty of options. You can start with one of the UI kits such as Bootstrap, Tailwind CSS, or Bulma as the base for your application UI. You can then build a design language or buy a pre-built, cheap UI theme from a theme market. If you are building an independent application, you can make your selections and create your UI pages and components based on these selections. Your pages and styling don't have to be compatible with another application.

On the other hand, if you want to build a modular application where each module's UI is independently developed (probably by a separate team) where the modules come together at runtime as a single application, you need to determine a design standard that should be implemented by all the module developers so that you have a consistent user interface.

Since ABP Framework provides a modular infrastructure, it provides a theming system that determines a set of base libraries and standards. This helps ensure that the application and module developers can build UI pages and components without depending on a particular theme or style set. Once the module/application code is theme-independent and the theme standards are explicit, you can build different themes and easily use that theme for an application with a simple configuration.

ABP Framework provides two free pre-built UI themes:

- The **Basic** theme is a minimalist theme that is built on the plain Bootstrap styling. It is ideal if you want to build styling from scratch.

- The **LeptonX** theme is a modern and production-ready UI theme built by the ABP Framework team.

This book uses the Basic theme in all examples. The following is a screenshot of the LeptonX theme:

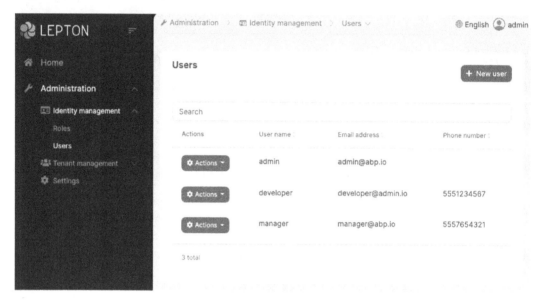

Figure 12.1 – The LeptonX theme and the application layout

Pre-built UI themes are deployed as NuGet and NPM packages, so you can easily install and switch between them.

The next two sections will introduce the fundamental base libraries and layouts that are shared by these themes.

The base libraries

To make modules/applications independent of a particular theme, ABP determines some base CSS and JavaScript libraries that our module/application can rely on.

The first and the most fundamental dependency of the ABP Framework MVC/Razor Pages UI is the *Twitter Bootstrap* framework. Starting with ABP Framework version 5.0, Bootstrap 5.x is used.

Besides Bootstrap, there are some other core library dependencies such as Datatables.Net, JQuery, JQuery Validation, FontAwesome, Sweetalert, Toastr, Lodash, and more. No additional setup is needed if you want to use these standard libraries in your module or application.

The next section will explain the layout system that is needed to understand how a web page is built.

The layouts

A typical web page consists of two parts – the layout and the page's content. The layout shapes the overall page and generally includes the main header, a company/product logo, the main navigation menu, a footer, and other standard components. The following screenshot shows these parts on an example layout:

Figure 12.2 – Parts of a page layout

In modern web applications, layouts are designed to be responsive, which means they change their shape and placing so that they are suitable for the device that's being used by the current user.

The layout's content almost remains the same across different pages – only the page's content changes. The page's content is generally a large part of the layout and may scroll if the content is larger than the height of the screen.

A web application may have different layout requirements in different parts/pages. In ABP Framework, a theme can define one or more layouts. Every layout has a unique `string` name, and ABP Framework defines four standard layout names:

- `Application`: Designed for back-office-style web applications with a header, menu, toolbar, footer, and so on. An example is shown in *Figure 12.1*.

- `Account`: Designed for login, register, and other account-related pages.

- **Public**: Designed for public-facing websites, such as a landing page for your product.

- **Empty**: A layout without an actual layout. The page's content covers the entire screen.

These strings are defined in the `Volo.Abp.AspNetCore.Mvc.UI.Theming.StandardLayouts` class. Every theme must define the `Application`, `Account`, and `Empty` layouts because they are common for most applications. The `Public` layout is optional and falls back to the `Application` layout if it's not implemented by the theme. A theme may define more layouts with different names.

The `Application` layout is the default unless you change it. You can change it per page/view or for a folder. If you change it for a folder, all the pages/views under that folder will use the selected layout.

To change it for a page/view, inject the `IThemeManager` service and use the `CurrentTheme.GetLayout` method with a layout name:

```
@inject IThemeManager ThemeManager
@{
    Layout = ThemeManager.CurrentTheme
            .GetLayout(StandardLayouts.Empty);
}
```

Here, you can use the `StandardLayouts` class to get the standard layout names. For this example, we could use `GetLayout("Empty")` since the value of `StandardLayouts.Empty` is a constant `string` that's `Empty`. In this way, you can get your theme's non-standard layouts with their `string` names.

If you want to change the layout for all the pages/views in a folder, you can create a `_ViewStart.cshtml` file in that folder and place the following code inside it:

```
@using Volo.Abp.AspNetCore.Mvc.UI.Theming
@inject IThemeManager ThemeManager
@{
    Layout = ThemeManager.CurrentTheme
            .GetLayout(StandardLayouts.Account);
}
```

If you place that `_ViewStart.cshtml` file in the `Pages` folder (or in `Views` for MVC views), all your pages will use the selected layout unless you select another layout for a subfolder or a particular page/view.

We can easily select a layout for our page to place content in. The next section will explain how to import script/style files into our pages and utilize the bundling and minification system.

Using the bundling and minification system

ABP offers an end-to-end solution for installing client-side packages, adding script/style files to the pages, and bundling and minifying these files in development and production environments.

Let's start by installing a client-side package for the application.

Installing NPM packages

NPM is the de facto package manager for JavaScript/CSS libraries. When you create a new solution with the MVC/Razor Pages UI, you will see a `package.json` file in the web project's root folder. The initial content of the `package.json` file will look something like this:

```
{
    ...
  "dependencies": {
      "@abp/aspnetcore.mvc.ui.theme.basic": "^5.0.0"
  }
}
```

Initially, we have a single NPM package dependency called `@abp/aspnetcore.mvc. ui.theme.basic`. This package has dependencies on all the base CSS/JavaScript libraries that are necessary for the Basic theme. If we want to install another NPM package, we can use the standard `npm install` (or `yarn add`) command.

Let's assume that we want to use the *Vue.js* library in our application. We can run the following command in the root directory of the web project:

```
npm install vue
```

This command installs the `vue` NPM package in the `node_modules/vue` folder. However, we can't use the files under the `node_modules` folder. We should copy the necessary files into the `wwwroot` folder of the web project to import them into the pages.

You can copy the necessary files manually, but this is not the best way. ABP provides an `install-libs` command to automate this process using a mapping file. Open the `abp.resourcemapping.js` file under the web project and add the following code to the `mappings` dictionary:

```
"@node_modules/vue/dist/vue.min.js": "@libs/vue/"
```

The final content of the `abp.resourcemapping.js` file should look as follows:

```
module.exports = {
    aliases: { },
    mappings: {
        "@node_modules/vue/dist/vue.min.js": "@libs/vue/"
    }
};
```

Now, we can run the following command in a command-line terminal, in the root directory of the web project:

```
abp install-libs
```

The `vue.min.js` file should be copied under the `wwwroot/libs/vue` folder:

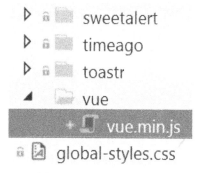

Figure 12.3 – Adding the Vue.js library to the web project

Mappings support glob/wildcard patterns. For example, you can copy all the files in the vue package with the following mapping:

```
"@node_modules/vue/dist/*": "@libs/vue/"
```

The libs folder is committed to the source control system (such as Git) by default. This means that if your teammate gets the code from your source control system, they don't need to run the npm install or abp install-libs commands. If you want, you can add the libs folder to the ignore file of your source control (such as .gitignore for Git). In this case, you need to run the npm install and abp install-libs commands before running the application.

The next section explains the standard ABP NPM packages.

Using the standard packages

Building a modular system has another challenge – all the modules should use the same (or compatible) version of the same NPM package. ABP Framework provides a set of standard NPM packages to allow the ABP ecosystem to use the same version of these NPM packages and automate the mapping to copy the resources to the libs folder.

@abp/vue is one of these standard packages that can be used to install the Vue.js library in your project. You can install this package instead of the vue package:

```
npm install @abp/vue
```

Now, you can run the abp install-libs command to copy the vue.min.js file into the wwwroot/libs/vue folder. Notice that you don't need to define the mapping in the abp.resourcemapping.js file since the @abp/vue package already includes the necessary mapping configuration.

It is suggested that you use the standard @abp/* packages when they are available. In this way, you can depend on a standard version of the related library, and you don't need to configure the abp.resourcemapping.js file manually.

However, when you install the library in your project, you will need to import it into the page to use it in your application.

Importing script and style files

Once we have installed a JavaScript or CSS library, we can include it in any page or bundle. Let's start with the most simple case – you can import vue.min.js into a Razor Page or view it using the following code:

```
@section scripts {
    <abp-script src=»/libs/vue/vue.min.js» />
}
```

Here, we are importing JavaScript files into the `scripts` section, so the theme is placing them at the end of the HTML document, after the global scripts. `abp-script` is a tag helper that's defined by ABP Framework to include scripts to the page/view. It is rendered as follows:

```
<script src=»/libs/vue/vue.min.js?_v=637653840922970000»></
script>
```

We could use a standard `script` tag, but `abp-script` has the following advantages:

- It automatically minifies the file in the production (or staging) environment if the given file is not already minified. If it is not minified and ABP finds the minified file near the original file, it uses the pre-minified file instead of dynamically minifying at runtime.

- It adds a query string parameter to add versioning information so that the browsers don't cache it when the file changes. This means that when you re-deploy your application, browsers don't accidentally cache the old versions of your script files.

- ABP ensures that the file is added to the page only once, even if you include it multiple times. This is a good feature if you wish to build a modular system since different module components may include the same library independent of each other, and ABP Framework eliminates this duplication.

Once we have included Vue.js in a page, we can utilize its power to create highly dynamic pages. Here is an example Razor Page, named `VueDemo.cshtml`:

```
@page
@model MvcDemo.Web.Pages.VueDemoModel
@section scripts {
    <abp-script src=»/libs/vue/vue.min.js» />
    <script>
        var app = new Vue({
            el: '#app',
            data: {
                message: 'Hello Vue!'
            }
        })
    </script>
}
<div id="app">
```

```
        <h2>{{ message }}</h2>
</div>
```

If you run this page, a **Hello Vue!** message will be shown on the UI. I can recommend using Vue.js in some pages of your MVC/Razor Pages applications when you need to build complex and dynamic user interfaces.

Let's take this example one step further and move the custom JavaScript code into a separate file. Create a JavaScript file named VueDemo.cshtml.js in the same folder:

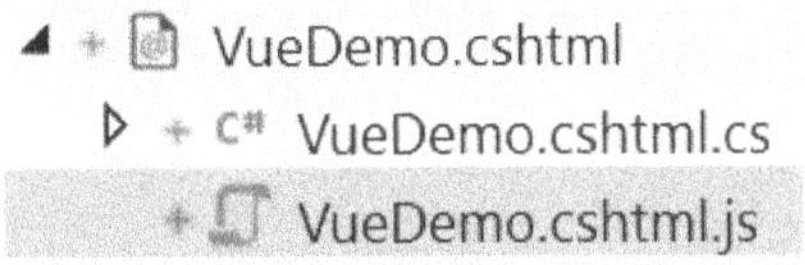

Figure 12.4 – Adding a JavaScript file

I prefer this naming convention, but you can set any name for the JavaScript file.

> **JavaScript/CSS Files Under the Pages Folder**
>
> In a regular ASP.NET Core application, you should place all the JavaScript/CSS files under the wwwroot folder. ABP allows you to add JavaScript/CSS files to the Pages or Views folder, near the corresponding .cshtml file. I find this approach to be pretty useful since we keep the related files together.

The content of the new JavaScript file is shown in the following code block:

```
var app = new Vue({
    el: '#app',
    data: {
        message: 'Hello Vue!'
    }
});
```

Now, we can update the VueDemo.cshtml file's content, as shown in the following code block:

```
@page
@model MvcDemo.Web.Pages.VueDemoModel
@section scripts {
    <abp-script src=»/libs/vue/vue.min.js» />
    <abp-script src=»/Pages/VueDemo.cshtml.js» />
```

```
}
<div id="app">
    <h2>{{ message }}</h2>
</div>
```

It is good to keep JavaScript code in a separate file and include it on the page as an external file, as in the preceding example.

Working with style (CSS) files is pretty similar to working with script files. The following example uses the `styles` section and the `abp-style` tag helper to import a style file on the page:

```
@section styles {
    <abp-style src="/Pages/VueDemo.cshtml.css" />
}
```

We can import multiple script or style files into a page. The next section will show you how to bundle these files as a single, minified file in production.

Creating page bundles

When we use multiple `abp-script` (or `abp-style`) tags on a page, ABP individually includes the files on the page and includes the minified versions in production. However, we generally want to create a single bundled and minified file in production. We can use the `abp-script-bundle` and `abp-style-bundle` tag helpers to create bundles for a page, as shown in the following example:

```
@section scripts {
    <abp-script-bundle>
        <abp-script src=»/libs/vue/vue.min.js» />
        <abp-script src=»/Pages/VueDemo.cshtml.js» />
    </abp-script-bundle>
}
```

Here, we are creating a bundle that includes two files. ABP automatically minifies these files and bundles them as a single file, and then versions this single file in the production environment. ABP makes the bundling operation in the first request to the page and then caches the bundled file in memory. It uses the cached bundle file for subsequent requests.

You can use conditional logic or dynamic code inside the bundle tags, as shown in the following example:

```
<abp-script-bundle>
    <abp-script src="/validator.js" />
    @if (System.Globalization.CultureInfo
        .CurrentUICulture.Name == "tr")
    {
        <abp-script src="/validator.tr.js" />
    }
    <abp-script src="/some-other.js" />
</abp-script-bundle>
```

This example adds a sample validation library to the bundle and conditionally adds the Turkish localization script. If the user's language is Turkish, then Turkish localization will be added to the bundle. Otherwise, it won't be added. ABP can understand the difference – it creates and caches two separate bundles, one for Turkish users and one for the rest.

With that, we've learned how to create bundles for an individual page. In the next section, we will explain how to configure global bundles.

Configuring global bundles

The bundling tag helpers are very useful for page bundles. You can also use them if you are creating custom layouts. However, when we use themes, the layouts are controlled by the theme.

Let's assume that we've decided to use the Vue.js library on all the pages and want to add it to the global bundle instead of adding it to every page individually. For this, we can configure AbpBundlingOptions in ConfigureServices of our module (in the web project), as shown in the following code block:

```
Configure<AbpBundlingOptions>(options =>
{
    options.ScriptBundles.Configure(
        StandardBundles.Scripts.Global,
```

```
        bundle =>
        {
            bundle.AddFiles(«/libs/vue/vue.min.js»);
        }
    );

    options.StyleBundles.Configure(
        StandardBundles.Styles.Global,
        bundle =>
        {
            bundle.AddFiles("/global-styles.css");
        }
    );
});
```

The `options.ScriptBundles.Configure` method is used to manipulate
a bundle with the given name. The first parameter is the name of the bundle.
`StandardBundles.Scripts.Global` is a `constant string` whose value is the
name of the global script bundle, which is imported by all the layouts. The preceding
example also adds a CSS file to the global style bundle.

The global bundles are just named bundles. We will explain these in the next section.

Creating named bundles

Page-based bundling is a simple way to create bundles for a single page. However, there
are situations where you will need to define a bundle and reuse it on multiple pages. As
explained in the previous section, the global style and script bundles were named bundles.
We can also define custom-named bundles and import the bundle in any page or layout.

The following example defines a named bundle and adds three JavaScript files inside it:

```
Configure<AbpBundlingOptions>(options =>
{
    options
        .ScriptBundles
        .Add("MyGlobalScripts", bundle => {
            bundle.AddFiles(
                "/libs/jquery/jquery.js",
                "/libs/bootstrap/js/bootstrap.js",
```

```
                "/scripts/my-global-scripts.js"
            );
        });
    });
```

We can write this code in `ConfigureServices` of a module class (typically, the module class in the web layer). `options.ScriptBundles` and `options.StyleBundles` are two kinds of bundles. In this example, we've used the `ScriptBundles` property to create a bundle that includes some JavaScript files.

Once we have created a named bundle, we can use it in a page/view using the `abp-script-bundle` and `abp-style-bundle` tag helpers, as shown in the following example:

```
<abp-script-bundle name="MyGlobalScripts" />
```

When we use this code in a page or view, all the script files are individually added to the page at development time. They are automatically bundled and minified in the production environment by default. The next section explains how to change this default behavior.

Controlling the bundling and minification behavior

We can use the `AbpBundlingOptions` options class to change the default behavior of the bundling and minification system. See the following configuration:

```
Configure<AbpBundlingOptions>(options =>
{
    options.Mode = BundlingMode.None;
});
```

This configuration code disables the bundling and minification logic. This means that even in production, all the script/style files are individually added to the page without bundling and minification. `options.Mode` can take one of the following values:

- `Auto` (default): Bundles and minifies in production and staging environments but disables bundling and minification at development time.

- `Bundle`: Bundles the files (creates a file per bundle) but does not minify the styles/scripts.

- `BundleAndMinify`: Always bundles and minifies the files, even at development time.

- `None`: Disables the bundling and minification process.

In this book, I've explained the basic usage of the bundling and minification system. However, it has advanced features, such as creating bundle contributor classes, inheriting a bundle from another bundle, extending and manipulating bundles, and more. These features are especially helpful when you want to create reusable UI modules. Please refer to the ABP Framework documentation for all the features: `https://docs.abp.io/en/abp/latest/UI/AspNetCore/Bundling-Minification`.

In the next section, you will learn how to work with navigation menus.

Working with menus

Menus are rendered by the current theme, so the final application or modules can't directly change the menu items. You can see the main menu on the left-hand side of *Figure 12.1*. ABP provides a menu system, so the modules and the final application can dynamically add new menu items or remove/change the items that are added by those modules.

We can use `AbpNavigationOptions` to add contributors to the menu system. ABP executes all the contributors to build the menu dynamically, as shown in the following example:

```
Configure<AbpNavigationOptions>(options =>
{
    options.MenuContributors.Add(new MyMenuContributor());
});
```

Here, `MyMenuContributor` should be a class that implements the `IMenuContributor` interface. The ABP startup solution template already contains a menu contributor class that you can directly use. `IMenuContributor` defines the `ConfigureMenuAsync` method, which we should implement like so:

```
public class MvcDemoMenuContributor : IMenuContributor
{
    public async Task ConfigureMenuAsync(
        MenuConfigurationContext context)
    {
        if (context.Menu.Name == StandardMenus.Main)
```

```
        {
            //TODO: Configure the main menu
        }
    }
}
```

The first thing we should consider is the menu's name. Two standard menu names are defined as constants in the StandardMenus class (in the Volo.Abp.UI.Navigation namespace):

- Main: The main menu of the application. It is shown on the left-hand side of *Figure 12.1*.

- User: The user context menu. It is opened when you click your username on the header.

So, the preceding example checks the menu's name and only adds items to the main menu. The following example code block adds a **Customer Relation Management (CRM)** menu item with two sub-menu items:

```
var l = context.GetLocalizer<MvcDemoResource>();
context.Menu.AddItem(
    new ApplicationMenuItem("MyProject.Crm", l["CRM"])
        .AddItem(new ApplicationMenuItem(
            name: "MyProject.Crm.Customers",
            displayName: l["Customers"],
            url: "/crm/customers")
        ).AddItem(new ApplicationMenuItem(
            name: "MyProject.Crm.Orders",
            displayName: l["Orders"],
            url: "/crm/orders")
        )
);
```

In this example, we are getting an IStringLocalizer instance (1) to localize the display names of the menu items. context.GetLocalizer is a shortcut to getting localizer services. You can use context.ServiceProvider to resolve any service and apply your custom logic to build the menu.

Every menu item should have a unique `name` (such as `MyProject.Crm.Customers` in this example) and a `displayName`. There are `url`, `icon`, `order`, and some other options available to control the appearance and behavior of the menu items.

The basic theme renders the example menu, as shown in the following screenshot:

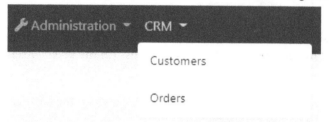

Figure 12.5 – Menu items rendered by the Basic theme

It is important to understand that the `ConfigureMenuAsync` method is called every time we render the menu. For a typical MVC/Razor Pages application, this method is called in every page request. In this way, you can dynamically shape the menu and conditionally add or remove items. You generally need to check permissions while adding the menu items, as shown in the following code block:

```
if (await context.IsGrantedAsync("MyPermissionName"))
{
    context.Menu.AddItem(...);
}
```

`context.IsGrantedAsync` is a shortcut for checking the permissions for the current user with a permission name. If we want to resolve and use `IAuthorizationService` manually, we could rewrite the same code, as shown in the following code block:

```
var authorizationService = context
    .ServiceProvider.
GetRequiredService<IAuthorizationService>();
if (await authorizationService.IsGrantedAsync(
    "MyPermissionName"))
{
    context.Menu.AddItem()
}
```

In this example, I used `context.ServiceProvider` to resolve `IauthorizationService`. Then, I used its `IsGrantedAsync` method, just like in the previous example. You can safely resolve services from `context.ServiceProvider` and let ABP Framework release these services at the end of the menu's build process.

It is also possible to find existing menu items (added by the depending modules) in the `context.Menu.Items` collection to modify or remove them.

In the next section, we will continue looking at Bootstrap tag helpers and learn how to render common Bootstrap components in a type-safe way.

Working with Bootstrap tag helpers

Bootstrap is one of the most popular UI (HTML/CSS/JS) libraries in the world, and it is the fundamental UI framework that's used by all the ABP themes. As a benefit of using such a library as a standard library, we can build our UI pages and components based on Bootstrap and let the theme style them. In this way, our modules and even applications can be theme-independent and work with any ABP-compatible UI theme.

Bootstrap is a well-documented and easy-to-use library. However, there are two problems while writing Bootstrap-based UI code:

- Some components require a lot of boilerplate code. Most parts of these codes are repetitive and tedious to write and maintain.
- Writing plain Bootstrap code in an MVC/Razor Pages web application is not very type-safe. We can make mistakes in class names and HTML structure that we can't catch at compile time.

ASP.NET Core MVC/Razor Pages has a *tag helper* system to define reusable components and use them as other HTML tags in our pages/views. ABP takes the power of tag helpers and provides a set of tag helper components for the Bootstrap library. In this way, we can build Bootstrap-based UI pages and components with less code and in a type-safe manner.

It is still possible to write native Bootstrap HTML code with ABP Framework, and ABP's Bootstrap tag helpers don't cover Bootstrap 100%. However, we suggest using the Bootstrap tag helpers wherever possible. See the following example:

```
<abp-button button-type="Primary" text="Click me!" />
```

Here, I used the abp-button tag helper to render a Bootstrap button. I used the button-type and text attributes with compile-time check support. This example code is rendered as follows at runtime:

```
<button class="btn btn-primary" type="button">
    Click me!
</button>
```

There are many Bootstrap tag helpers in ABP Framework, so I won't explain all of them here. Please refer to ABP's documentation to learn how to use them: https://docs.abp.io/en/abp/latest/UI/AspNetCore/Tag-Helpers/Index.

In the next two sections, we will use some of these Bootstrap tag helpers to build form items and open modals.

Creating forms and implementing validation

ASP.NET Core provides a good infrastructure for preparing forms and submitting, validating, and processing them on the server side. However, it still requires writing some boilerplate and repeating code. ABP Framework simplifies working with forms by providing tag helpers and automating validation and localization wherever possible. Let's begin with how to render form elements using ABP's tag helpers.

Rendering form elements

The abp-input tag helper is used to render an appropriate HTML input element for a given property. It is better to show its usage in a complete example.

Let's assume that we need to build a form to create a new *movie* entity and have created a new Razor Page called CreateMovie.cshtml. First, let's look at the code-behind file:

```
public class CreateMovieModel : AbpPageModel
{
    [BindProperty]
    public MovieViewModel Movie { get; set; }

    public void OnGet()
    {
        Movie = new MovieViewModel();
    }
}
```

```
    public async Task OnPostAsync()
    {
        // TODO: process the form (using the Movie object)
    }
}
```

Page models are normally derived from the `PageModel` class. However, we are deriving from ABP's `AbpPageModel` base class since it provides some pre-injected services and helper methods. That's a simple page model class. Here, we are creating a new `MovieViewModel` instance in the `OnGet` method to bind it to the form elements. We also have an `OnPostAsync` method that we can use to process the posted form data. `[BindProperty]` tells ASP.NET Core to bind the post data to the `Movie` object.

To explore this example, let's look at the `MovieViewModel` class:

```
public class MovieViewModel
{
    [Required]
    [StringLength(256)]
    public string Name { get; set; }

    [Required]
    [DataType(DataType.Date)]
    public DateTime ReleaseDate { get; set; }

    [Required]
    [TextArea]
    [StringLength(1000)]
    public string Description { get; set; }

    public Genre Genre { get; set; }
    public float? Price { get; set; }
    public bool PreOrder { get; set; }
}
```

This object is used to render the form elements and bind the post data when the user submits the form. Notice that some properties have data annotation validation attributes to validate the values of these properties automatically. Here, the `Genre` property is an enum, as shown here:

```
public enum Genre
{
    Classic, Action, Fiction, Fantasy, Animation
}
```

Now, we can switch to the view part and try to render a form to get the movie information from the user.

First, I will you show how we can do this without ABP Framework to understand the benefits of using ABP Framework. First, we must open a `form` element, as shown in the following code block:

```
<form method="post">
    <-- TODO: FORM ELEMENTS -->
    <button class="btn btn-primary" type="submit">
        Submit
    </button>
</form>
```

In the `form` block, we write code for each `form` element, and then we add a `submit` button to post the form. Showing the full code of `form` would be too long for this book, so I will only show the code that is necessary for rendering the input element for the `Movie.Name` property:

```
<div class="form-group">
    <label asp-for="Movie.Name" class="control-label">
    </label>
    <input asp-for="Movie.Name" class="form-control"/>
    <span asp-validation-for="Movie.Name"
        class="text-danger"></span>
</div>
```

The preceding code block should seem very familiar to you if you have ever created a form with ASP.NET Core Razor Pages/MVC and Bootstrap. It puts a `label`, the actual input element, and a validation message area by wrapping them with a `form-group`. The following screenshot shows the rendered form:

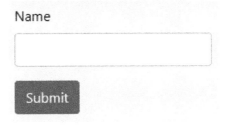

Figure 12.6 – A simple form with a single text input

The form currently contains only a single piece of text input for the `Name` property. You could write similar code for each property of the `Movie` class, which would result in large and repetitive code. Let's see how we can render the same input using ABP Framework's `abp-input` tag helper:

```
<abp-input asp-for="Movie.Name" />
```

That's pretty easy. Now, we can render all the form elements. The following is the finalized code:

```
<form method="post">
    <abp-input asp-for="Movie.Name" />
    <abp-select asp-for="Movie.Genre" />
    <abp-input asp-for="Movie.Description" />
    <abp-input asp-for="Movie.Price" />
    <abp-input asp-for="Movie.ReleaseDate" />
    <abp-input asp-for="Movie.PreOrder" />
    <abp-button type="submit" button-type="Primary"
        text="Submit"/>
</form>
```

The preceding code block is dramatically shorter compared to the standard Bootstrap form code. I used the `abp-select` tag helper for the `Genre` property. It understands that `Genre` is an enum and creates the dropdown element using the enum members. The following is the rendered form:

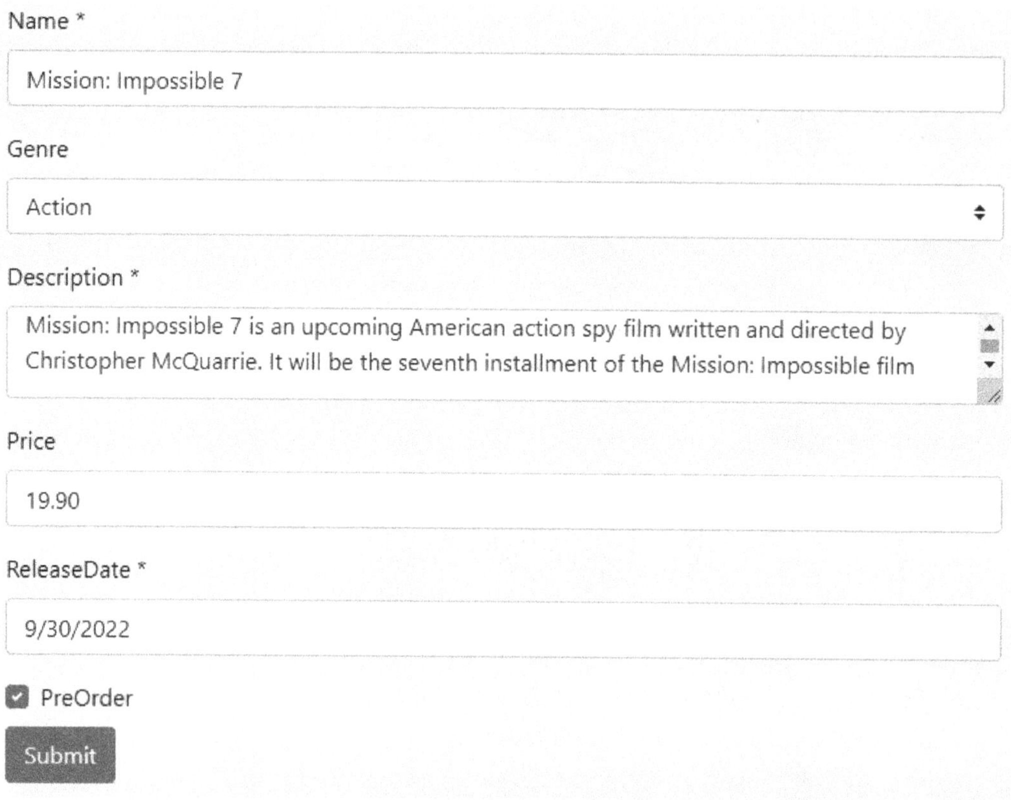

Name *

> Mission: Impossible 7

Genre

> Action   ⬍

Description *

> Mission: Impossible 7 is an upcoming American action spy film written and directed by
> Christopher McQuarrie. It will be the seventh installment of the Mission: Impossible film

Price

> 19.90

ReleaseDate *

> 9/30/2022

☑ PreOrder

Submit

Figure 12.7 – Full form for creating a new movie

ABP automatically adds * near the label of required form fields. It reads the types and attributes of the class properties and determines the form fields.

If all you want to do is render the input elements in order, you can replace the last code block with the following one:

```
<abp-dynamic-form abp-model="Movie" submit-button="true" />
```

The `abp-dynamic-form` tag helper gets a model and creates the entire form automatically!

The abp-input, abp-select, and abp-radio tag helpers are mapped to a class property and render the corresponding UI element. You can use them if you want to control the layout of the form and place custom HTML elements between the form controls. On the other hand, abp-dynamic-form makes creating the form super simple while you have less control over the form's layout. However you create the form, ABP automates the validation and localization process for you, as I will explain in the next few sections.

Validating user inputs

If you try to submit the form without filling in the required fields, the form won't be submitted to the server, and an error message will be shown for each invalid form element. The following screenshot shows the error message when you leave the Name property empty and submit the form:

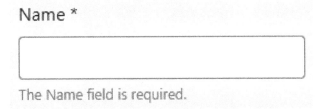

Figure 12.8 – Invalid user input

Client-side validation is automatically done based on the data annotation attributes in the MovieViewModel.Name property. So, you don't need to write any validation code for the standard checks. Users can't submit the form until all the fields are valid.

Client-side validation is just for the user experience. It would be easy to bypass the client-side validation and submit an invalid form to the server (by manipulating or disabling the JavaScript code in the browser's developer tools). So, you should always validate the user input on the server side, which should be done in the OnPostAsync method of the page model class. The following code block shows the common pattern that's used while handling a form post:

```
public async Task OnPostAsync()
{
    if (ModelState.IsValid)
    {
        //TODO: Create a new movie
    }
    else
```

```
    {
        Alerts.Danger("Please correct the form fields!");
    }
}
```

`ModelState.IsValid` returns `false` if any form field is invalid. This is a standard feature of ASP.NET Core. You should always process the input in such an `if` statement. Optionally, you can have logic in the `else` statement. In this example, I used ABP's `Alerts` feature to show a client-side alert message to the user. The following screenshot shows the result of submitting the invalid form:

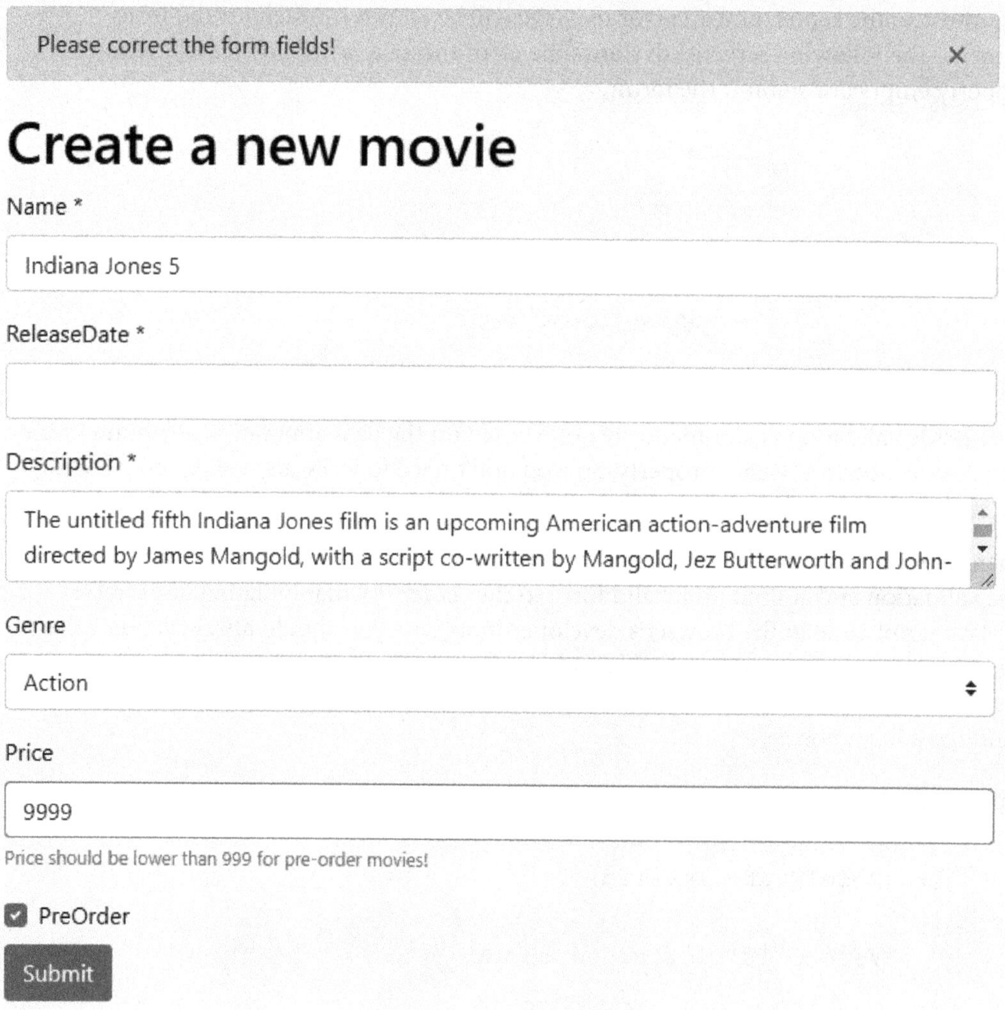

Figure 12.9 – Invalid form result from the server

If you look at the validation error message under the **Price** field, you will see a custom error message. I've implemented the `IValidatableObject` interface for the `MovieViewModel` class, as shown in the following code block:

```
public class MovieViewModel : IValidatableObject
{
    // ... properties omitted
    public IEnumerable<ValidationResult> Validate(
        ValidationContext validationContext)
    {
        if (PreOrder && Price > 999)
        {
            yield return new ValidationResult(
                "Price should be lower than 999 for
                pre-order movies!",
                new[] { nameof(Price) }
            );
        }
    }
}
```

I'm performing complex custom validation logic in the `Validate` method. You can refer to the *Validating user inputs* section in *Chapter 7, Exploring Cross-Cutting Concerns*, to learn more about server-side validation. Here, we should understand that we can use custom logic on the server and show validation messages on the client side.

In the next section, we will learn how to localize validation errors, as well as form labels.

Localizing forms

ABP Framework automatically localizes the validation error messages based on the current language. Try to switch to another language and submit the form without providing a movie name. The following screenshot shows this for the Turkish language:

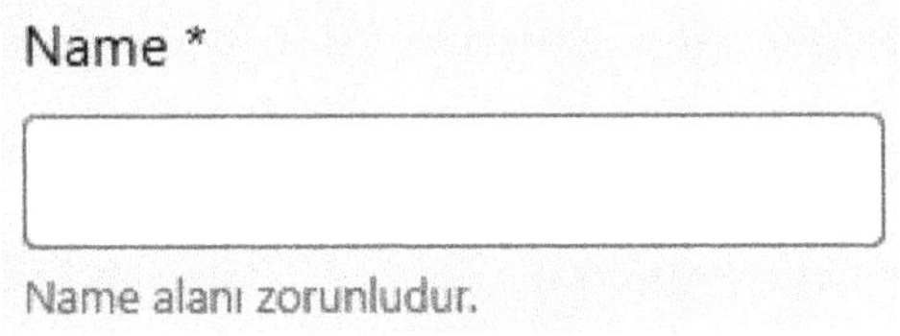

Name *

Name alanı zorunludur.

Figure 12.10 – Auto-localized validation error messages

The error text has changed. However, you can still see **Name** as the field name because that's our custom field name, and we haven't localized it yet.

ABP provides a convention-based localization system for the form fields. You just define a localization entry in your localization JSON file with the key formatted as `DisplayName:<property-name>`. I can add the following lines to the `en.json` file (in the *Domain.Shared* project) to localize all the fields of the movie creation form:

```
"DisplayName:Name": "Name",
"DisplayName:ReleaseDate": "Release date",
«DisplayName:Description»: «Description»,
«DisplayName:Genre»: «Genre»,
"DisplayName:Price": "Price",
"DisplayName:PreOrder": "Pre-order"
```

Then, I can localize these in the Turkish language with the following entries in the `tr.json` file:

```
"DisplayName:Name": "İsim",
"DisplayName:ReleaseDate": "Yayınlanma tarihi",
"DisplayName:Description": "Açıklama",
"DisplayName:Genre": "Tür",
"DisplayName:Price": "Ücret",
"DisplayName:PreOrder": "Ön sipariş"
```

Now, we have a localized label and a more localized validation error message:

İsim *

İsim alanı zorunludur.

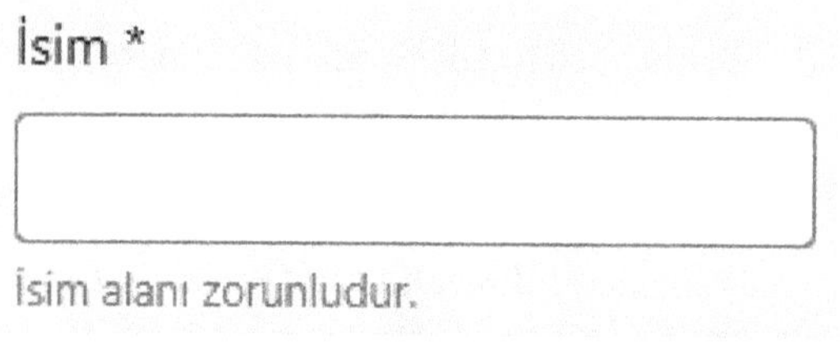

Figure 12.11 – Fully localized validation error message and the field label

Adding the `DisplayName:` prefix to the property name is a suggested convention for the `form` fields, but actually, it is not required. If ABP can't find the `DisplayName:Price` entry, it will search an entry with the `Price` key, without any prefix. If you want to specify the localization key for a property, you can add the `[DisplayName]` attribute on top of the property, as shown in the following example:

```
[DisplayName("MoviePrice")]
public float? Price { get; set; }
```

With this setup, ABP will try to localize the field name using the `"MoviePrice"` key.

The `abp-select` tag localizes the items in the dropdown for `enum` types by conventions. You can add entries to your localization file, such as `<enum-type>.<enum-member>`. For the `Action` member of the `Genre` enum type, we can add a localization entry with the `Genre.Action` key. It falls back to the `Action` key if the `Genre.Action` key is not found.

In the next section, we will discuss how to convert a standard form into a fully AJAX form.

Implementing AJAX forms

When the user submits a standard form, a full-page post is performed and the server re-renders the entire page. An alternative approach could be posting the form as an AJAX request and handling the response in JavaScript code. This approach is much faster than the regular post request since the browser doesn't need to reload the whole page and all the resources of the page. It is also a better user experience in many cases since you can show some animations on the waiting time. Also, in this way, you don't lose the page's state and can perform smart actions in your JavaScript code.

You can handle all the AJAX stuff manually, but ABP Framework provides built-in ways for such common patterns. You can add the `data-ajaxForm="true"` attribute to any `form` element (including the `abp-dynamic-form` element) to make it posted through an AJAX request.

The following example adds the AJAX feature to `abp-dynamic-form`:

```
<abp-dynamic-form abp-model="Movie"
                  submit-button="true"
                  data-ajaxForm="true"
                  id="MovieForm" />
```

When we convert a form into an AJAX form, the post handler on the server side should be implemented properly. The following code block shows a common pattern to implement the post handler:

```
public async Task<IActionResult> OnPostAsync()
{

    ValidateModel();
    //TODO: Create a new movie
    return NoContent();

}
```

The first line validates the user input and throws AbpValidationException if the input model is not valid. The ValidateModel method comes from the base AbpPageModel class. If you don't want to use it, you can check if (ModelState. IsValid) and take any action you need. If the form is valid, you normally save the new movie to a database. Finally, you can return the resulting data to the client. We don't need to return a response for this example, so a NoContent result is fine.

When you convert a form into an AJAX form, you typically want to take action when the form is successfully submitted. The following example handles the abp-ajax-success event of the form:

```
$(function () {
    $('#MovieForm').on('abp-ajax-success', function(){
        $('#MovieForm').slideUp();
        abp.message.success('Successfully saved, thanks
            :)');
    });
});
```

In this example, I registered a callback function for the abp-ajax-success event of the form. In this callback, you can do anything you need. For example, I used the slideUp JQuery function to hide the form, then used ABP's success UI message. We will return to the abp.message API in the *Using the JavaScript API* section of this chapter.

Exception handling logic is different for AJAX requests. ABP handles all exceptions, returns a proper JSON response to the client, and then automatically handles the error on the client side. For example, suppose the form has a validation error that's been determined on the server side. In this case, the server returns a validation error message, and the client shows a message box, as shown in the following screenshot:

Your request is not valid!

The following errors were detected during validation.
- Price should be lower than 999 for pre-order movies!

Figure 12.12 – Server-side validation errors on the AJAX form submit

The message box is shown in any exception, including your custom exceptions and
`UserFriendlyException`. Go to the *Exception handling* section of *Chapter 7*,
Exploring Cross-Cutting Concerns, to learn more about the exception handling system.

In addition to converting the form into an AJAX form and handling exceptions, ABP also
prevents double-clicking on the **Submit** button to prevent multiple posts. The **Submit**
button becomes busy, and the button's text changes to **Processing...** until the request is
completed. You can set the `data-busy-text` attribute on the **Submit** button to use
another piece of text.

In the next section, we will learn how ABP Framework helps us while working with
modal dialogs.

Working with modals

A modal is one of the essential components when you want to create interactive user
interfaces. It provides a convenient way to get a response from the user or show some
information without changing the current page layout.

Bootstrap has a modal component, but it requires some boilerplate code. ABP Framework
provides the `abp-modal` tag helper to render a modal component, which simplifies the
modal's usage in most use cases. Another problem with modals is placing the modal code
inside the page that opens the modal, which makes the modal hard to reuse. ABP provides
a modal API on the JavaScript side to dynamically load and control these modals. It also
works well with forms inside modals. Let's begin with the simplest usage.

Understanding the basics of modals

ABP suggests defining modals as separate Razor Pages (or views if you are using the MVC pattern). So, as the first step, we should create a new Razor Page. Let's assume that we've created a new Razor Page called `MySimpleModal.cshtml` under the `Pages` folder. The code-behind file is simple:

```
public class MySimpleModalModel : AbpPageModel
{
    public string Message { get; set; }

    public void OnGet()
    {
        Message = "Hello modals!";
    }
}
```

We just have a `Message` property being shown inside the modal dialog. Let's see the view side:

```
@page
@model MvcDemo.Web.Pages.MySimpleModalModel
@{
    Layout = null;
}
<abp-modal>
    <abp-modal-header title="My header"></abp-modal-header>
    <abp-modal-body>
        @Model.Message
    </abp-modal-body>
    <abp-modal-footer buttons="Close"></abp-modal-footer>
</abp-modal>
```

The `Layout = null` statement is critical here. Because this page is loaded with an AJAX request, the result should only contain the modal's content, not the standard layout. `abp-modal` is the main tag helper that renders the HTML of the modal dialog. `abp-modal-header`, `abp-modal-body`, and `abp-modal-footer` are the main parts of the modal and have different options. The modal body is very simple in this example; it just shows `Message` on the model.

We've created the modal, but we should create a way to open it. ABP provides the `ModalManager` API on the JavaScript side to control a modal. Here, we need to create a `ModalManager` object on the page where we want to open the modal:

```
var simpleModal = new abp.ModalManager({
    viewUrl: '/MySimpleModal'
});
```

`abp.ModalManager` has a few options, but the most basic is `viewUrl`, which indicates the URL where the modal content will be loaded. Once we have a `ModalManager` instance, we can call its `open` method to open the modal:

```
$(function (){
    $('#Button1').click(function (){
        simpleModal.open();
    });
});
```

This example assumes there is a button with an ID of `Button1` on the page. We are opening the modal when the user clicks the button. The following screenshot shows the opened modal:

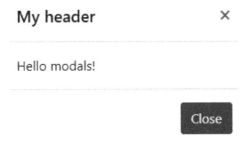

Figure 12.13 – A simple modal dialog box

Typically, we create dynamic content in a modal, so we need to pass some arguments while opening the modal dialog box. To do this, you can pass an object to the open method that contains the modal arguments, as shown in the following example:

```
simpleModal.open({
    productId: 42
});
```

Here, we passed a productId argument to the modal, so it may show details of the given product. You can add the same argument to the OnGet method of the MySimpleModalModel class to obtain the value and process inside the method:

```
public void OnGet(int productId)
{
    ...
}
```

You can get the product information from the database and render the product details in the modal body.

In the next section, we will learn how to place a form inside a modal to get data from the user.

Working with forms inside modals

Modals are widely used to show a form to users. ABP's ModalManager API gracefully handles some common tasks for you:

- It focuses on the first input of the form.

- It triggers a validation check when you press the *Enter* key or click the **Save** button. It doesn't allow you to submit the form unless the form is fully valid.

- It submits the form via an AJAX request, disables the modal buttons, and shows a progressing icon until the save operation is complete.

- If you've entered some data and click the **Cancel** button or close the modal, it warns you about unsaved changes.

Let's assume that we want to show a modal dialog to create a new movie and we've created a new Razor Page called `ModalWithForm.cshtml`. The code-behind file is similar to what we had in the *Implementing AJAX forms* section:

```
public class ModalWithForm : AbpPageModel
{
    [BindProperty]
    public MovieViewModel Movie { get; set; }

    public void OnGet()
    {
        Movie = new MovieViewModel();
    }

    public async Task<IActionResult> OnPostAsync()
    {
        ValidateModel();
        //TODO: Create a new movie
        return NoContent();
    }
}
```

The `OnPostAsync` method validates the user input first. If the form is not valid, an exception is thrown and handled by ABP Framework on the server side and the client side. You can return a response to the client, but we are returning a `NoContent` response in this example.

The view side of the modal is a bit different since we mix a form with a modal:

```
@page
@using Volo.Abp.AspNetCore.Mvc.UI.Bootstrap.TagHelpers.Modal
@model MvcDemo.Web.Pages.ModalWithForm
@{
    Layout = null;
}
<form method="post" asp-page="/ModalWithForm">
    <abp-modal>
        <abp-modal-header title="Create new movie">
        </abp-modal-header>
```

```
        <abp-modal-body>
            <abp-input asp-for="Movie.Name" />
            <abp-select asp-for="Movie.Genre" />
            <abp-input asp-for="Movie.Description" />
            <abp-input asp-for="Movie.Price" />
            <abp-input asp-for="Movie.ReleaseDate" />
            <abp-input asp-for="Movie.PreOrder" />
        </abp-modal-body>
        <abp-modal-footer buttons="@(
            AbpModalButtons.Cancel|AbpModalButtons.Save)">
        </abp-modal-footer>
    </abp-modal>
</form>
```

The abp-modal tag is wrapped by a form element. We don't put the form tag inside the abp-modal-body element because the **Save** button (which submits the form) in the modal footer should be inside form. So, as a solution, we are placing form as the topmost element in this view. The rest of the code block should be familiar; we use ABP input tag helpers to render the form elements.

Now, we can open the modal in our JavaScript code:

```
var newMovieModal = new abp.ModalManager({
    viewUrl: '/ModalWithForm'
});
$(function (){
    $('#Button2').click(function (){
        newMovieModal.open();
    });
});
```

The opened dialog is shown in the following screenshot:

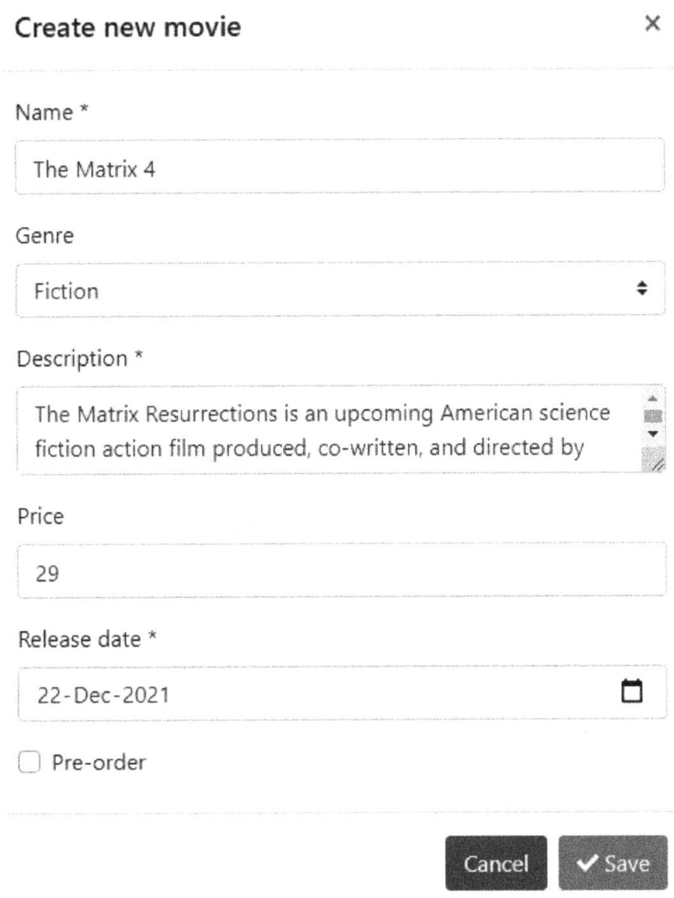

Figure 12.14 – A form inside a modal

It is also possible to use the `abp-dynamic-form` tag helper within a modal. We could rewrite the modal's view like so:

```
<abp-dynamic-form abp-model="Movie"
    asp-page="ModalWithForm">
    <abp-modal>
        <abp-modal-header title="Create new movie!">
        </abp-modal-header>
        <abp-modal-body>
            <abp-form-content/>
        </abp-modal-body>
        <abp-modal-footer buttons="@(
```

```
                        AbpModalButtons.Cancel|AbpModalButtons.Save)">
        </abp-modal-footer>
    </abp-modal>
</abp-dynamic-form>
```

Here, I wrapped `abp-modal` with an `abp-dynamic-form` element, just like in the previous section. The main point of this example is that I used the `<abp-form-content/>` tag helper in the `abp-modal-body` element. `abp-form-content` is an optional tag helper that is used to place the form inputs of the `abp-dynamic-form` tag helper in the desired place.

You typically want to take action once the modal form has been saved. For this, you can register a callback function to the `onResult` event of `ModalManager`, as shown in the following code block:

```
newMovieModal.onResult(function (e, data){
    console.log(data.responseText);
});
```

`data.responseText` will be the data if the server sends any result. For example, you can return a `Content` response from the `OnPostAsync` method, as shown in the following example:

```
public async Task<IActionResult> OnPostAsync()
{
    ...
    return Content("42");
}
```

ABP simplifies all these common tasks. Otherwise, you would write a lot of boilerplate code.

In the next section, we will learn how to add client-side logic to our modal dialogs.

Adding JavaScript for modals

If your modal needs some advanced client-side logic, you may want to write some custom JavaScript code for your modal. You can write your JavaScript code on the page where you open the modal, but that is not very modular and reusable. It is good to write your modal's JavaScript code in a separate file, ideally near the `.cshtml` file of the modal (remember that ABP allows you to place JavaScript files under the `Pages` folder).

For this, we can create a new JavaScript file and define a function in the `abp.modals` namespace, as shown in the following code:

```
abp.modals.MovieCreation = function () {
    this.initModal = function(modalManager, args) {
        var $modal = modalManager.getModal();
        var preOrderCheckbox =
            $modal.find('input[name="Movie.PreOrder"]');
        preOrderCheckbox.change(function(){
            if (this.checked){
                alert('checked pre-order!');
            }
        });
        console.log('initialized the modal...');
    }
};
```

Once we have created such a JavaScript class, we can associate it with the modal while creating the `ModalManager` instance:

```
var newMovieModal = new abp.ModalManager({
    viewUrl: '/ModalWithForm',
    modalClass: 'MovieCreation'
});
```

`ModalManager` creates a new instance of the `abp.modals.MovieCreation` class for every time you open the modal and calls the `initModal` function if you define it. The `initModal` function takes two parameters. The first one is the `ModalManager` instance that's associated with the modal so that you can use its functions. The second parameter is the arguments that you passed to the `open` function while opening the modal.

The `initModal` function is a perfect place to prepare the modal's content and register some callbacks to the events of the modal components. In the preceding example, I got the modal instance and a JQuery object, found the `Movie.PreOrder` checkbox, and registered its `change` callback so that I'm informed when the user checks it.

This example still doesn't work yet since we haven't added the JavaScript file to the page. There are two ways to add it to the page:

- We can use the `abp-script` tag to include the modal's JavaScript file in the page where we open the modal.

- We can set up `ModalManager` so that it lazy loads the JavaScript file.

The first option is straightforward – just include the following line in the page where you want to use the modal:

```
<abp-script src="/Pages/ModalWithForm.cshtml.js" />
```

If we want to lazy load the modal's script, we can configure `ModalManager` like so:

```
var newMovieModal = new abp.ModalManager({
    viewUrl: '/ModalWithForm',
    scriptUrl: '/Pages/ModalWithForm.cshtml.js',
    modalClass: 'MovieCreation'
});
```

Here, I added the `scriptUrl` option as the URL of the modal's JavaScript file. `ModalManager` lazy loads the JavaScript file the first time you open the modal. The script is not loaded again if you open the modal a second time (without refreshing the whole page).

In this section, we learned how to work with forms, validation, and modals. They are essential parts of a typical web application. In the next section, we will learn about some useful JavaScript APIs that we need in every application.

Using the JavaScript API

In this section, we will explore some useful client-side APIs of ABP Framework. Some of these APIs provide simple ways to use server-side defined features such as authentication and localization, while others provide solutions for common UI patterns, such as message boxes and notifications.

All the client-side JavaScript APIs are global objects and functions that are declared under the `abp` namespace. Let's begin with accessing the current user's information in your JavaScript code.

Accessing the current user

We are using the ICurrentUser service on the server side to get information about the currently logged-in user. In the JavaScript code, we can use the global abp. currentUser object, as shown here:

```
var userId = abp.currentUser.id;
var userName = abp.currentUser. userName;
```

By doing this, we can get the user's ID and username. The following JSON object is an example of the abp.currentUser object:

```
{
    isAuthenticated: true,
    id: "813108d7-7108-4ab2-b828-f3c28bbcd8e0",
    tenantId: null,
    userName: "john",
    name: "John",
    surName: "Nash",
    email: "john.nash@abp.io",
    emailVerified: true,
    phoneNumber: "+901112223342",
    phoneNumberVerified: true,
    roles: ["moderator","manager"]
}
```

If the current user has not logged in yet, all these values will be null or false, as you would expect. The abp. currentUser object provides an easy way to get information about the current user. In the next section, we will learn how to check the permissions of the current user.

Checking user permissions

ABP's authorization and permission management system is a powerful way to define permissions and check them at runtime for the current user. Checking these permissions in your JavaScript code is effortless using the abp. auth API.

The following example checks if the current user has the `DeleteProduct` permission:

```
if (abp.auth.isGranted('DeleteProduct')) {
  // TODO: Delete the product
} else {
  abp.message.warn("You don't have permission to delete
                products!");
}
```

`abp.auth.isGranted` returns `true` if the current user has given permission or a policy. If the user doesn't have permission, we show a warning message using the ABP message API, which will be explained in the *Showing message boxes* section later in this chapter.

While these APIs are rarely needed, you can use the `abp.auth.policies` object when you need to get a list of all the available permissions/policies and the `abp.auth.grantedPolicies` object if you need to get a list of all the granted permissions/policies for the current user.

Hiding UI Parts Based on Permissions

A typical use case for client-side permission checking is to hide some UI parts (such as action buttons) based on the user's permissions. While the `abp.auth` API provides a dynamic way to do that, I suggest using the standard `IAuthorizationService` on your Razor Pages/views to conditionally render the UI elements wherever possible.

Note that checking permissions on the client side is just for the user experience and that it doesn't guarantee security. You should always check the same permission on the server side.

In the next section, we will learn how to check the feature rights of the current tenant in a multi-tenant application.

Checking the tenant features

The feature system is used to restrict application functionalities/features based on the current tenant. We will explore ABP's multi-tenancy infrastructure in *Chapter 16, Implementing Multi-Tenancy*. However, we will cover checking tenant features here for the ASP.NET Core MVC/Razor Pages UI.

The `abp.features` API is used to check feature values for the current tenant. Let's assume that we have a feature for importing email lists from Mailchimp (a cloud email marketing platform) and that we've defined a feature named `MailchimpImport`. We can easily check if the current tenant has that feature enabled:

```
if (abp.features.isEnabled('MailchimpImport'))
{
    // TODO: Import from Mailchimp
}
```

`abp.features.isEnabled` only returns `true` if the given feature's value is `true`. ABP's feature system allows you to define non-boolean features too. In this case, you can use the `abp.features.get (...)` function to obtain the given feature's value for the current tenant.

Checking features on the client side makes it easy to perform dynamic client-side logic, but remember to check the features on the server side as well for a secure application.

In the next section, we will continue using the localization system in your JavaScript code.

Localizing strings

One powerful part of ABP's localization system is that you can reuse the same localization strings on the client side. In this way, you don't have to deal with another kind of localization library in your JavaScript code.

The `abp.localization` API is available in your JavaScript code to help you utilize the localization system. Let's begin with the simplest case:

```
var str = abp.localization.localize('HelloWorld');
```

The `localize` function, with that usage, takes a localization key and returns the localized value based on the current language. It uses the default localization resource. If you need to, you can specify the localization resource as the second parameter:

```
var str = abp.localization.localize('HelloWorld',
 'MyResource');
```

Here, we've specified `MyResource` as the localization resource. If you want to localize lots of strings from the same resource, there is a shorter way to do this:

```
var localizer = abp.localization.getResource('MyResource');
var str = localizer('HelloWorld');
```

Here, you can use the `localizer` object to get texts from the same resource.

The JavaScript localization API applies the same fallback logic to the server-side API; it returns the given key if it can't find the localized value.

If the localized string contains placeholders, you can pass the placeholder values as parameters. Let's assume that we have the following entry in the localization JSON file:

```
"GreetingMessage": "Hello {0}!"
```

We can pass a parameter to the `localizer` or `abp.localization.localize` function, as shown in the following example:

```
var str = abp.localization.localize('GreetingMessage', 'John');
```

The resulting `str` value will be `Hello John!` for this example. If you have more than one placeholder, you can pass the values to the `localizer` function in the same order.

Besides the localizing texts, you may need to know the current culture and language so that you can take extra actions. The `abp.localization.currentCulture` object contains detailed information about the current language and culture. In addition to the current language, the `abp.localization.languages` value is an array of all the available languages in the current application. Most of the time, you don't directly use these APIs since the theme you're using is responsible for showing a list of languages to the user and allows you to switch between them. However, it is good to know that you can access the language data when you need it.

So far, you've learned how to use some ABP server-side features on the client side. In the next section, you will learn how to show message and confirmation boxes to the user.

Showing message boxes

It is very common to show blocking message boxes to users to inform them about something important happening in the application. In this section, you will learn how to show nice message boxes and confirmation dialogs in your applications.

The `abp.message` API is used to show a message box to inform the user easily. There are four types of message boxes:

- `abp.message.info`: Displays an informative message

- `abp.message.success`: Displays a success message

- `abp.message.warn`: Displays a warning message

- `abp.message.error`: Displays an error message

Let's take a look at the following example:

```
abp.message.success('Your changes have been successfully
                    saved!', 'Congratulations');
```

In this example, I've used the `success` function to display a success message. The first parameter is a message text, while the optional second parameter is a message header. The result of this example is shown in the following screenshot:

Congratulations

Your changes have been successfully saved!

Figure 12.15 – A success message box

Message boxes are blocked, which means the page is blocked (non-clickable) until the user clicks the **OK** button.

Another kind of message box is used for confirmation purposes. The `abp.message.confirm` function shows some dialog to get a response from the user:

```
abp.message.confirm('Are you sure to delete this product?')
.then(function(confirmed){
  if(confirmed){
    // TODO: Delete the product!
  }
});
```

The confirm function returns a promise, so we could chain it with the then callback to execute some code once the user closes the dialog by accepting or canceling it. The following screenshot shows the confirmation dialog that was created for this example:

Are you sure?

Are you sure to delete this product?

Figure 12.16 – A confirmation dialog

Message boxes are a good way to grab a user's attention. However, there is an alternative way to do this, as we'll see in the next section.

Showing notifications

Notifications are a non-blocking way to inform users of certain events. They are shown on the bottom right corner of the screen and automatically disappear after a few seconds. Just like the message boxes, there are four types of notifications:

- abp.notify.info: Displays an informative notification
- abp.notify.success: Displays a success notification
- abp.notify.warn: Displays a warning notification
- abp.notify.error: Displays an error notification

The following example shows an information notification:

```
abp.notify.info(
    'The product has been successfully deleted.',
    'Deleted the Product'
);
```

The second parameter is the notification title and is optional. The result of this example code is shown in the following screenshot:

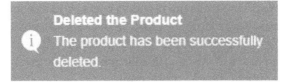

Figure 12.17 – A notification message

With the notification API, we are closing the JavaScript APIs.

Here, I covered the most used APIs. However, there are more APIs you can use in your JavaScript code, all of which you can learn about by reading the ABP Framework documentation: `https://docs.abp.io/en/abp/latest/UI/AspNetCore/ JavaScript-API/Index`. In the next section, we will learn how to consume server-side APIs from JavaScript code.

Consuming HTTP APIs

You can use any tool or technique to consume HTTP APIs from your JavaScript code. However, ABP provides the following ways as fully integrated solutions:

- You can use the `abp.ajax` API as an extension of the `jQuery.ajax` API.

- You can use dynamic JavaScript client proxies to call server-side APIs, just like you can with JavaScript functions.

- You can generate static JavaScript client proxies at development time.

Let's begin with the first one – the `abp.ajax` API.

Using the abp.ajax API

The `abp.ajax` API is a wrapper around the standard `jQuery.ajax` API. It automatically handles all errors and shows a localized message to the user on an error case. It also adds the anti-forgery token to the HTTP header to satisfy **Cross-Site Request Forgery** (**CSRF**) protection on the server side.

The following example uses the `abp.ajax` API to get a list of users from the server:

```
abp.ajax({
  type: 'GET',
  url: '/api/identity/users'
}).then(function(result){
  // TODO: process the result
});
```

In this example, we've specified GET as the request's type. You can specify all the standard options of jQuery.ajax (or $.ajax) to override the defaults. abp.ajax returns a promise object, so we could add the then callback to handle the result that's sent by the server. We can also use the catch callback to handle errors and the always callback to perform an action at the end of the request.

The following example shows how to handle errors manually:

```
abp.ajax({
    type: 'GET',
    url: '/api/identity/users',
    abpHandleError: false
}).then(function(result){
    // TODO: process the result
}).catch(function(){
    abp.message.error("request failed :(");
});
```

Here, I added a catch callback function after the then function. You can perform your error logic here. I also specified the abpHandleError: false option to disable ABP's automatic error handling logic. Otherwise, ABP will handle the error and show an error message to the user.

abp.ajax is a low-level API. You typically use dynamic or static client proxies to consume your own HTTP APIs.

Using dynamic client proxies

You should have already used the dynamic JavaScript client proxy system if you applied the example application from *Chapter 3, Step-By-Step Application Development*. ABP Framework generates JavaScript functions at runtime to easily consume all the HTTP APIs of your application.

The following code block shows two sample methods of IProductAppService that were defined in *Chapter 3, Step-By-Step Application Development*:

```
namespace ProductManagement.Products
{
    public interface IProductAppService :
        IApplicationService
    {
```

```
        Task CreateAsync(CreateUpdateProductDto input);
        Task<ProductDto> GetAsync(Guid id);
    }
}
```

All of these methods are available in the same namespace on the client side. For example, we can get a product by its ID, as shown in the following code block:

```
productManagement.products.product
    .get('1b8517c8-2c08-5016-bca8-39fef5c4f817')
    .then(function (result) {
      console.log(result);
    });
```

`productManagement.products` is the camel case equivalent of the `ProductManagement.Products` namespace of the C# code. `product` is the conventional name of `IProductAppService`. The `I` prefix and the `AppService` suffix have been removed, and the remaining name is converted into camel case. Then, we can use the method name that's been converted in camel case without the `Async` suffix. So, the `GetAsync` method is used as the `get` function in the JavaScript code. The `get` function takes the same parameters that the C# method takes. It returns a `Deferred` object so that we can chain it with the `then`, `catch`, or `always` callbacks, similar to what can do for the `abp.ajax` API. It internally uses the `abp.ajax` API. In this example, the `result` argument of the `then` function is the `ProductDto` object that's sent by the server.

Other methods are used in a similar way. For example, we can create a new product with the following code:

```
productManagement.products.product.create({
    categoryId: '5f568193-91b2-17de-21f3-39fef5c4f808',
    name: 'My product',
    price: 42,
    isFreeCargo: true,
    releaseDate: '2023-05-24',
    stockState: 'PreOrder'
});
```

Here, we pass the `CreateUpdateProductDto` object with the JSON object format.

In some cases, we may need to pass additional AJAX options for the HTTP API calls. You can pass an object as the last parameter to every proxy function:

```
productManagement.products.product.create({
    categoryId: '5f568193-91b2-17de-21f3-39fef5c4f808',
    name: 'My product',
    //...other values
}, {
    url: 'https://localhost:21322/api/my-custom-url'
    headers: {
        'MyHeader': 'MyValue'
    }
});
```

Here, I passed an object to change the URL and add a custom header to the request. You can refer to jQuery's documentation (https://api.jquery.com/jquery.ajax/) for all the available options.

Dynamic JavaScript client proxy functions are generated at runtime by the /Abp/ ServiceProxyScript endpoint of your application. This URL is added to the layout by the themes so that you can directly use any proxy function in your pages without importing any script.

In the next section, you will learn about an alternative way to consume your HTTP APIs.

Using static client proxies

Unlike dynamic client proxies, which are generated at runtime, static proxies are generated at development time. We can use the ABP CLI to generate the proxy script file.

First, we need to run the application that serves the HTTP APIs because the API endpoint data is requested from the server. Then, we can use the generate-proxy command, as shown in the following example:

```
abp generate-proxy -t js -u https://localhost:44349
```

The generate-proxy command can take the following parameters:

- -t (required): The type of the proxy. We use js for JavaScript here.
- -u (required): The root URL of the API endpoint.

- -m (optional): The module name to generate the proxy for. The default value is app and is used to generate proxies for your application. In a modular application, you can specify the module name here.

Static JavaScript proxies are generated under the wwwroot/client-proxies folder, as shown in the following screenshot:

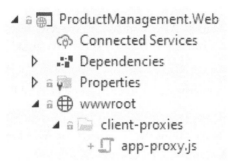

Figure 12.18 – The static JavaScript proxy file

Then, you can import the proxy script file into any page and use the static proxy functions like you would the dynamic ones.

When you use static proxies, you don't need dynamic proxies. By default, ABP creates dynamic proxies for your application. You can configure DynamicJavaScriptProxyOptions to disable it for the application, as shown in the following example:

```
Configure<DynamicJavaScriptProxyOptions>(options => {
    options.EnabledModules.Remove("app");
});
```

The EnabledModules list contains app by default. If you are building a modular application and want to enable dynamic JavaScript proxies for your module, you need to add it to the EnabledModules list explicitly.

Summary

In this chapter, we covered the fundamental design points and the essential features of the MVC/Razor Pages UI of ABP Framework.

The theming system allows you to build theme/style independent modules and applications and easily switch between UI themes. It makes this possible by defining a set of base libraries and standard layouts.

You then learned about the bundling and minification system, which covers the entire development cycle of importing and using client-side dependencies in your applications and optimizing resource usage in a production environment.

ABP makes it easy to create forms and implement validation and localization using tag helpers and predefined conventions. You also learned how to convert a standard form into an AJAX-submitted form.

We've also covered some JavaScript APIs that can utilize ABP features on the client side, such as authorization and localization, and easily show nice-looking message boxes and notifications.

Finally, you learned about alternative ways to consume HTTP APIs from your JavaScript code.

In the next chapter, you will learn about the Blazor UI for ABP Framework to build interactive web UIs using C# instead of JavaScript.

13
Working with the Blazor WebAssembly UI

Blazor is a relatively new **Single-Page Application (SPA)** framework for building interactive web applications using C# instead of JavaScript. Blazor is one of the built-in UI options provided by ABP Framework.

In this chapter, I will briefly discuss what Blazor is and the main pros and cons of using this new framework. I will then continue by explaining how you can create new ABP solutions using the Blazor UI option. At the end of the chapter, you will have understood the architecture and design of the ABP Blazor integration and learned about the essential ABP services that you will use in your applications.

This chapter consists of the following topics:

- What is Blazor?
- Getting started with the ABP Blazor UI
- Authenticating the user
- Understanding the theming system

- Working with menus

- Using the basic services

- Using the UI services

- Consuming HTTP APIs

- Working with global scripts and styles

Technical requirements

If you want to follow the examples in this chapter, you need to have an IDE/editor that supports ASP.NET Core development. We will use the ABP CLI at some points, so you need to install the ABP CLI, as explained in *Chapter 2, Getting Started with ABP Framework*.

You can download the example application from the following GitHub repository: `https://github.com/PacktPublishing/Mastering-ABP-Framework`. It contains some of the examples given in this chapter.

What is Blazor?

As I indicated in the introduction, Blazor is an SPA framework used to build interactive web applications, just like other SPA frameworks such as Angular, React, and Vue.js. However, it has one important difference – we can use C# to build the application instead of JavaScript, which means we can run .NET in browsers. Blazor uses the .NET core runtime to execute the .NET code in the browser (for Blazor WebAssembly).

Running .NET in browsers is not a new idea. Microsoft has done it before with Silverlight. To run Silverlight applications, we had to install a plugin on the browser. Blazor, on the other hand, runs natively on the browser, thanks to **WebAssembly** technology, which is defined as the following on `https://webassembly.org`:

> *"WebAssembly (abbreviated Wasm) is a binary instruction format for a stack-based virtual machine. Wasm is designed as a portable compilation target for programming languages, enabling deployment on the web for client and server applications."*

A higher-level language, such as C#, can be compiled into WebAssembly and run natively in the browser. WebAssembly is supported by all major web browsers, so we don't need to install any custom plugin. If you're wondering whether Blazor is the new Silverlight, I can simply say, no, it is not.

As .NET developers, Blazor brings incredible opportunities to us:

- We can use our existing C# skills to develop applications by harnessing the full power of the language and the runtime.

- We can use existing .NET libraries, such as our favorite NuGet packages.

- We can share code (such as DTO classes, application service contracts, localization, and validation code) between the server and the client.

- We can use the familiar Razor syntax to build UI pages and components.

Besides using C#, Blazor provides JavaScript interoperability to call JavaScript code from C# and vice versa. That means that we can use existing JavaScript libraries and write our JavaScript code whenever we need to.

Writing C# and sharing code between server and client applications is a huge advantage for a .NET developer. ABP also takes advantage of this and shares the infrastructure between the MVC/Razor Pages UI and Blazor UI as much as possible. You will see that many services are very similar to the MVC/Razor Pages UI.

As a .NET developer and a software company manager, I am very impressed by Blazor and will use it in future projects. However, that doesn't mean it has no drawbacks:

- The bundle size, initial load time, and runtime performance are worse than its JavaScript competitors, such as Angular and React. However, Microsoft is investing in Blazor and working hard to improve its performance. For example, **Ahead-of-Time** (**AOT**) compilation has been introduced with .NET 6.0.

- The UI components and ecosystem are not mature yet since Blazor is still in the early stages.

- Debugging is not so straightforward yet.

If these drawbacks are tolerable for your projects, you can definitely start using Blazor today.

Interestingly, Blazor has two kinds of runtime models. Until now, I have mostly talked about **Blazor WebAssembly**. The second model is called **Blazor Server**. While the component development model is identical, the hosting logic and the runtime model are completely different.

With Blazor WebAssembly, .NET code runs in the browser on the Mono runtime, and we don't have to run .NET on the server side. A small initializer JavaScript code downloads the standard .NET **Dynamic Link Libraries** (**DLLs**) and runs them in the browser. This model is similar to, and a direct competitor of, Angular and React because it runs the client-side logic completely in the browser.

On the other hand, Blazor Server runs .NET code completely on the server. It establishes a real-time SignalR connection between the client and the server. The browser runs JavaScript and communicates to the server over that SignalR connection. It sends events to the server, and the server executes the necessary .NET code and sends **Document Object Model (DOM)** changes to the browser. Finally, the browser applies the DOM changes to the UI.

The Blazor Server model has a pretty faster initial load time compared to Blazor WebAssembly. However, it communicates to the server for all events and DOM changes, so we need a good and stable connection between the server and the client.

My purpose in this book is not to provide a complete introduction, overview, and use cases of Blazor but to give a short enough introduction to understanding what it is. Also, this chapter will focus on Blazor WebAssembly, but most of the topics are applicable to Blazor Server.

Now, we can start ABP's Blazor integration.

Getting started with the ABP Blazor UI

There are two ways to start a new project using ABP's startup solution templates. You can either download it from `https://abp.io/get-started` or create it using the ABP CLI. I will use the CLI approach in this book. If you haven't installed it yet, open a command-line terminal and execute the following command:

```
dotnet tool install -g Volo.Abp.Cli
```

Now, we can create a new solution using the `abp new` command:

```
abp new DemoApp -u blazor
```

`DemoApp` is the solution name in this example. I've passed the `-u blazor` parameter to specify Blazor WebAssembly. If you want to use Blazor Server, you can specify the parameter as `-u blazor-server`.

I haven't specified a database provider, so it uses Entity Framework Core by default (specify the `-d mongodb` parameter if you want to use MongoDB). After creating the solution, we need to create the initial database migration. As a first step, we should execute the following command in the `src/DemoApp.DbMigrator` directory:

```
dotnet run
```

This command creates the initial code-first migration and applies against the database.

The solution contains two applications:

- The first one is the server (backend) application that hosts the HTTP APIs and provides the authentication UI.

- The second application is the frontend Blazor WebAssembly application that contains the application UI and communicates to the server.

So, we first run the `DemoApp.HttpApi.Host` server application for this example. Then, we can run the `DemoApp.Blazor` Blazor application to run the UI. You can click on the **Login** link, and type `admin` as the username and `1q2w3E*` as the password to log in to the application.

I won't dig into the details of the application, since we've done it already in *Chapter 2, Getting Started with ABP Framework*. The next section explains how the user is authenticated.

Authenticating the user

OpenID Connect (**OIDC**) is Microsoft's suggested way to authenticate Blazor WebAssembly applications. ABP follows that suggestion and provides it as preconfigured in the startup solution.

The Blazor application doesn't contain login, register, or other authentication-related UI pages. It uses the **Authorization Code** flow with **Proof Key for Code Exchange** (**PKCE**) enabled to redirect the user to the server application. The server handles all the authentication logic and redirects the user back to the Blazor application.

The authentication configuration is stored in the `wwwroot/appsettings.json` file of the Blazor application. See the following example configuration:

```
"AuthServer": {
  "Authority": "https://localhost:44306",
  "ClientId": "DemoApp_Blazor",
  "ResponseType": "code"
}
```

Here, `Authority` is the backend server application's root URL. `ClientId` is the name of the Blazor application that is known by the server. Finally, `ResponseType` specifies the authorization code flow.

This configuration is used in the module class, the `DemoAppBlazorModule` class for this example, as shown in the following code block:

```
private static void ConfigureAuthentication(
    WebAssemblyHostBuilder builder)
{
    builder.Services.AddOidcAuthentication(options =>
    {
        builder.Configuration.Bind(
            "AuthServer", options.ProviderOptions);
        options.UserOptions.RoleClaim = JwtClaimTypes.Role;
        options.ProviderOptions.DefaultScopes.Add(
            "DemoApp");
        options.ProviderOptions.DefaultScopes.Add("role");
        options.ProviderOptions.DefaultScopes.Add("email");
        options.ProviderOptions.DefaultScopes.Add("phone");
    });
}
```

`AuthServer` is the key that matches the configuration key. If you want to customize the authentication options, these are the points you need to start from. For example, you can revise the requested scopes or change the OIDC configuration. For more information about Blazor WebAssembly authentication, please refer to Microsoft's documentation: `https://docs.microsoft.com/en-us/aspnet/core/blazor/security/webassembly/`.

In the next section, I will introduce the theming system for the Blazor UI.

Understanding the theming system

ABP provides a theming system for the Blazor UI, as explained when we covered the MVC/Razor Pages UI in *Chapter 12, Working with MVC/Razor Pages*. The theme system brings flexibility, so we can develop our applications and modules without depending on a particular UI theme/style.

All of the ABP themes for the Blazor UI use a set of base libraries. The fundamental base library is Bootstrap, whose components are designed to work with JavaScript. Fortunately, some component libraries wrap the Bootstrap components and provide a simpler .NET API, which is more suitable for use in Blazor applications.

One of these component libraries is **Blazorise**. It is actually an abstraction library and can work with multiple providers such as Bootstrap, Bulma, and Ant Design. ABP startup templates use the Bootstrap provider of the Blazorise library.

You can learn more about Blazorise and see the components in action on its website: `https://blazorise.com`. The following figure is a screenshot from the form components demo:

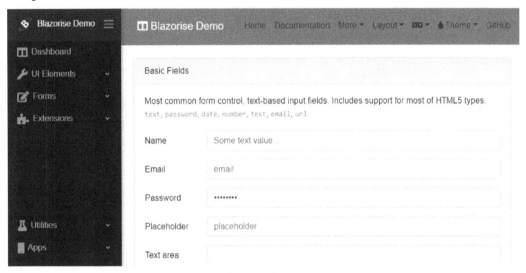

Figure 13.1 – Blazorise demo: form components

Besides the Blazorise library, the ABP Blazor UI uses **Font Awesome** as the CSS font icon library. So, any module or application can use these libraries on their pages without an explicit dependency.

The UI theme is responsible for rendering the layout, including the header, menu, toolbar, page alerts, and footer.

In the next section, we will see how to add new items to the main menu.

Working with menus

Menu management in the ABP Blazor UI is very similar to the ABP MVC/Razor Pages UI, which was covered in *Chapter 12, Working with MVC/Razor Pages*.

We use `AbpNavigationOptions` to add contributors to the menu system. ABP executes all the contributors to build the menu dynamically. The startup solution includes a menu contributor and is added to `AbpNavigationOptions` as per the following example:

```
Configure<AbpNavigationOptions>(options =>
{
    options.MenuContributors.Add(new
        DemoAppMenuContributor(
        context.Services.GetConfiguration()));
});
```

`DemoAppMenuContributor` is a class that implements the `IMenuContributor` interface. The `IMenuContributor` interface defines the `ConfigureMenuAsync` method, which we should implement as shown in the following example:

```
public class DemoAppMenuContributor : IMenuContributor
{
    public async Task ConfigureMenuAsync(
        MenuConfigurationContext context)
    {
        if (context.Menu.Name == StandardMenus.Main)
        {
            //TODO: Configure the main menu
        }
    }
}
```

There are two standard menu names defined as constants in the `StandardMenus` class (in the `Volo.Abp.UI.Navigation` namespace):

- `Main`: The main menu of the application.
- `User`: The user context menu. It is opened when you click your username on the header.

So, the preceding example checks the menu name and adds items only to the main menu. The following code block adds a new menu item to the main menu:

```
var l = context.GetLocalizer<DemoAppResource>();
context.Menu.AddItem(
```

```
    new ApplicationMenuItem(
        DemoAppMenus.Home,
        l["Menu:Home"],
        "/home",
        icon: "fas fa-home"
    )
);
```

You can resolve services from dependency injection using the context.
ServiceProvider object. The context.GetLocalizer method is a shortcut to
resolve an IStringLocalizer<T> instance. Similarly, we can use the context.
IsGrantedAsync shortcut method to check the permission of the current user, as
shown in the following code block:

```
if (await context.IsGrantedAsync("MyPermissionName"))
{
    context.Menu.AddItem(...);
}
```

Menu items can be nested. The following example adds a Crm menu item and an Orders
menu item under it:

```
context.Menu.AddItem(
    new ApplicationMenuItem(
        DemoAppMenus.Crm,
        l["Menu:Identity"]
    ).AddItem(new ApplicationMenuItem(
        DemoAppMenus.Orders,
        l["Menu:Orders"],
        url: "/crm/orders")
    )
);
```

I've called AddItem on the first ApplicationMenuItem object to add a child item.
You can do the same for the Orders menu item to build deeper menus.

We've used localization and authorization services while creating menu items. In the next
section, we will see how to use these services in other parts of our Blazor application.

Using the basic services

In this section, I will show you how to use some fundamental services in Blazor applications. As you will see, they are almost the same as the server-side services that we covered in earlier chapters. Let's start with the authorization service.

Authorizing the users

We typically use authorization in Blazor applications to hide/disable some pages, components, and functionalities on the user interface. While the server always checks the same authorization rules for security, client-side authorization checks provide a better user experience.

IAuthorizationService is used to programmatically check permissions/policies, as on the server side. You can inject and use its methods, as shown in the following example:

```
public partial class Index
{
    protected override async Task OnInitializedAsync()
    {
        if (await AuthorizationService
                .IsGrantedAsync("MyPermission"))
        {
            // TODO: ...
        }
    }
}
```

AuthorizationService has different ways to work. Please refer to the *Working with authorization and permission systems* section of *Chapter 7, Exploring Cross-Cutting Concerns*, to learn more about the authorization system.

The component in the preceding example is inherited from the AbpComponentBase class. We can directly use the AuthorizationService property without manual injection since the AbpComponentBase class pre-injects it for us. The AuthorizationService property type is IAuthorizationService.

If you don't inherit from the `AbpComponentBase` class, you can inject it using the `[Inject]` attribute:

```
[Inject]
private IAuthorizationService AuthorizationService { get;
                                                     set; }
```

You can use the same `IAuthorizationService` on the view side of your Razor components when you need it. However, there are some alternative ways to make your application code cleaner. For example, you can use the `[Authorize]` attribute on a component to make it available only for authenticated users:

```
@page "/"
@attribute [Authorize]
<p>This page is visible only if you've logged in</p>.
```

The `[Authorize]` attribute works similarly to the server side. You can pass a policy/permission name to check for a specific permission, as shown in the following example:

```
@page "/order-management"
@attribute [Authorize("CanManageOrders")]
<p>You can only see this if you have the necessary
    permission.</p>
```

It is typical to show a part of the UI if the user has a specific permission. The following example uses the `AuthorizeView` element to show a message if the current user has permission to edit orders:

```
<AuthorizeView Policy="CanEditOrders">
    <p>You can only see this if you can edit the
        orders.</p>
</AuthorizeView>
```

In this way, you can conditionally render the action buttons or other parts of the UI.

ABP is 100% compatible with Blazor's authorization system, so you can refer to Microsoft's documentation to see more examples and details: `https://docs.microsoft.com/en-us/aspnet/core/blazor/security`.

In the next section, we will learn how to use the localization system, another common UI service.

Localizing the user interface

Blazor applications share the same API for localizing texts. We can inject and use the `IStringLocalizer<T>` service to get the localized texts for the current language.

The following Razor component uses the `IStringLocalizer<T>` service:

```
@using DemoApp.Localization
@using Microsoft.Extensions.Localization
@inject IStringLocalizer<DemoAppResource> L
<h3>@L["HelloWorld"]</h3>
```

We use the standard `@inject` directive and specify the localization resource type in the generic `IStringLocalizer<T>` interface. The same interface can also be injected and used in any service in your application. Please refer to the *Localizing the user interface* section of *Chapter 8*, *Using the Features and Services of ABP*, to learn about working with the localization system.

Next, we will learn to get information about the current user in the next section.

Accessing the current user

You sometimes may need to know the current user's username, email address, and other details in your application. We use the `ICurrentUser` service to access the current user, as on the server side. The following example component renders a welcome message by the current user's name:

```
@using Volo.Abp.Users
@inject ICurrentUser CurrentUser
<h3>Welcome @CurrentUser.Name</h3>
```

In addition to standard properties such as `Name`, `Surname`, `UserName`, and `Email`, you can use the `ICurrentUser.FindClaimValue(...)` method to get custom claims issued by the server.

I've introduced the basic ABP Blazor services typically used by all the applications. I kept them short since the APIs are almost the same with the server side, and we've already covered them in detail in previous chapters. In the next section, I will continue with the UI services used to inform the user.

Using the UI services

It is common in every application to show messages, notifications, and alerts to users to inform or warn them. In the next sections, I will introduce ABP's built-in APIs for these services.

Showing message boxes

Message boxes are used to show blocking messages or confirmation dialogs to the user. The user clicks on the **Ok** button to disable the message or clicks the **Yes** or **Cancel** buttons to make a decision on configuration dialogs.

There are five types of messages – Info, Success, Warn, Error, and Confirm. The following example shows a Success message sent to the user:

```
@page "/"
@inherits DemoAppComponentBase
<Button Color="Color.Primary"
        Clicked="ShowSuccess">Click me!</Button>
@code
{
    private async Task ShowSuccess()
    {
        await Message.Success("This is a success
                               message!");
    }
}
```

The Message property, in this example, is coming from the AbpComponentBase class (DemoAppComponentBase inherits it), and its type is IUiMessageService. Alternatively, you can inject IUiMessageService manually for your components, pages, or services. All the IUiMessageService methods can take an extra title parameter and an options action to customize the dialog.

The following figure shows the result of the preceding example:

This is a success message!

Figure 13.2 – A simple success message without a title

The following example shows a confirmation dialog sent to the user and takes action if the user clicks the **Yes** button:

```
@page "/"
@inherits DemoAppComponentBase
<Button Color="Color.Primary"
        Clicked="ShowQuestion">Click me!</Button>
@code
{
    private async Task ShowQuestion()
    {
        var result = await Message.Confirm(
            "Are you sure to delete the product?");
        if (result == true)
        {
            //TODO: ...
        }
    }
}
```

The Confirm method returns a bool value, so you can see whether the user has accepted the dialog message. The following figure shows the result of this example:

Are you sure to delete the product?

Figure 13.3 – A confirmation dialog

The next section explains how to show a non-blocking information message to the user.

Showing notifications

Messages boxes focus users on the message. They should click the **Ok** button to return to the application UI. On the other hand, notifications are non-blocking informative messages. They are shown in the bottom-right corner of the screen and automatically disappear after a few seconds.

There are four types of notifications – Info, Success, Warn, and Error. The following example shows a confirmation dialog and shows a Success notification if the user accepts the confirmation message:

```
@page "/"
@inherits DemoAppComponentBase
<Button Color="Color.Primary"
        Clicked="ShowQuestion">Click me!</Button>
@code
{
    private async Task ShowQuestion()
    {
        var confirmed = await Message.Confirm(
            "Are you sure to delete the product?");
        if (confirmed)
```

```
        {
            //TODO: Delete the product
            await Notify.Success("Successfully deleted the
                               product!");
        }
    }
}
```

The `Notify` property comes from the `AbpComponentBase` base class. You can inject the `IUiNotificationService` interface and use it anywhere to show notifications on the UI. All the notification methods can take an extra `title` parameter and an `options` action to customize the dialog. The following figure shows the result of the `Notify.Success` method used in the preceding code block:

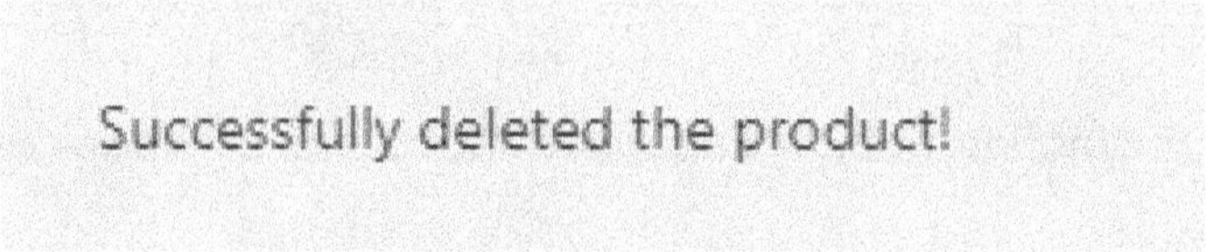

Successfully deleted the product!

Figure 13.4 – An example notification message

The next section introduces alerts, another way to show a message to the user.

Showing alerts

Using alerts is a sticky way to show a non-blocking message to the user. The user, optionally, can dismiss the alert.

There are four types of alerts – `Info`, `Success`, `Warning`, and `Danger`. The following example shows a `Success` alert sent to the user:

```
@page "/"
@inherits DemoAppComponentBase
<Button Color="Color.Primary"
        Clicked="DeleteProduct">Click me!</Button>
@code
{
    private async Task DeleteProduct()
    {
        //TODO: Delete the product
        Alerts.Success(
```

```
            text: "Successfully deleted the product.",
            title: "Deleted!",
            dismissible: true);
    }
}
```

In this example, I've used the `Alerts` property coming from the base class. You can always inject the `IAlertManager` service and use it like `IAlertManager.Alerts.Success(…)`.

All of the alert methods take `text` (required), a `title` (optional), and `dismissible` (optional and default – `true`) parameters. If an alert is dismissible, then the user can make it disappear by clicking the **X** button. The following figure shows the alert created in the preceding example:

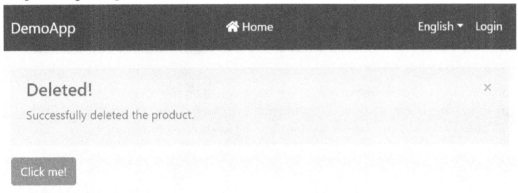

Figure 13.5 – A Success alert message

Alerts are rendered by the theme on top of the page content. Besides the standard `Info`, `Success`, `Warning`, and `Danger` methods, you can use the `Add` method by specifying `AlertType` to use all Bootstrap styles, such as `Primary`, `Secondary`, or `Dark`.

You've now learned three ways to show information messages to the user. In the next section, we will investigate how the Blazor application consumes the HTTP APIs of the server.

Consuming HTTP APIs

You can use the standard `HttpClient` to manually set up and perform an HTTP request to the server. However, ABP provides C# client proxies to call HTTP API endpoints easily. You can directly consume your application services from the Blazor UI and let ABP Framework handle the HTTP API calls for you.

Let's assume that we have an application service interface, as shown in the following example:

```
public interface ITestAppService : IApplicationService
{
    Task<int> GetDataAsync();
}
```

Application service interfaces are defined in the `Application.Contracts` project (the `DemoApp.Application.Contracts` project for the example solution I've created). The Blazor application has a reference to that project. This way, we can use the `ITestAppService` interface on the client side.

Application services are implemented in the `Application` project (the `DemoApp.Application` project for the example solution I've created). We can simply implement the `ITestAppService` interface, as shown in the following code block:

```
public class TestAppService : ApplicationService,
    ITestAppService
{
    public async Task<int> GetDataAsync()
    {
        return 42;
    }
}
```

Now, we can directly inject `ITestAppService` into any page/component, as with any other local service, and call its methods, just like a standard method call:

```
public partial class Index
{
    [Inject]
    private ITestAppService TestAppService { get; set; }
```

```
    private int Value { get; set; }

    protected override async Task OnInitializedAsync()
    {
        Value = await TestAppService.GetDataAsync();
    }
}
```

In this example, I used the standard `[Inject]` attribute on top of the
`TestAppService` property to tell Blazor to inject it for me. Then, I've overridden it in
the `OnInitializedAsync` method to call the `GetDataAsync` method. As we know,
the `OnInitializedAsync` method is called just after the component/page is initially
rendered and ready to work.

It's that easy. When we call the `GetDataAsync` method, ABP actually makes an
HTTP API call to the server by handling all the complexity, including authentication,
error handling, and JSON serialization. It reads the server's root URL from the
`RemoteServices` configuration in the `wwwroot/appsettings.json` file of the
Blazor project. An example configuration is shown in the following code block:

```
"RemoteServices": {
  "Default": {
    "BaseUrl": "https://localhost:44306"
  }
}
```

In this section, I've used ABP's dynamic C# client proxy approach to consume HTTP APIs
from the Blazor application. We will return to this topic in *Chapter 14, Building HTTP
APIs and Real-Time Services*, by also introducing the static C# client proxies.

The next section will explore how we can add script and style files to our Blazor
applications.

Working with global scripts and styles

Importing script and style files for the Blazor Server UI is the same as the MVC/Razor
Pages UI for ABP Framework. You can refer to *Chapter 12, Working with MVC/Razor
Pages*, to learn how to use it. This section is based on Blazor WebAssembly.

Blazor WebAssembly is a Single-Page Application and it has a single entry point by default. The `index.html` file is in the `wwwroot` folder, as shown in the following figure:

Figure 13.6 – The index.html file in the wwwroot folder

`index.html` is a plain HTML file. The server sends it to the browser without any processing. Remember that a simple static file server can serve a Blazor WebAssembly application. The browser first loads the `index.html` document and then loads the styles and scripts imported by this document.

If you open the `index.html` document, you will see a part within the `ABP:Styles` comments, as shown in the following code block:

```
<!--ABP:Styles-->
<link href="global.css?_v=637649661149948696"
    rel="stylesheet"/>
<link href="main.css" rel="stylesheet"/>
<!--/ABP:Styles-->
```

This code part (including the comments) is automatically created (and then updated) by the ABP CLI when you execute the following command in the root folder of the Blazor project:

```
abp bundle
```

When you execute this command, it creates (or regenerates) the global style bundle. This bundle contains all the necessary styles, including the .NET runtime, Blazor, and other used libraries, in a minified format. Whenever you add a new Blazor-related ABP NuGet package/module into your application, you rerun the `abp bundle` command and it regenerates the bundle with the necessary dependencies included.

ABP's `bundle` command does a great job. When installing a module, you don't need to know its global script files or extra dependencies. Just run this command, and you have the updated, production-ready global bundle file. Every module contributes its own dependencies into that bundle, and then ABP generates the bundle by respecting the module dependency order. To manipulate the bundle, you should define a class that implements the `IBundleContributor` interface. The Blazor project in the startup solution template already contains a bundle contributor, as shown in the following code block:

```
public class DemoAppBundleContributor : IBundleContributor
{
    public void AddScripts(BundleContext context)
    {
    }

    public void AddStyles(BundleContext context)
    {
        context.Add("main.css", excludeFromBundle: true);
    }
}
```

`AddScripts` and `AddStyles` methods are used to add JavaScript and CSS files to the global bundles. You can also remove or change an existing file (which was added by a package your application depends on) using the `context.BundleDefinitions` collection, but that's rarely needed. Here, the `excludeFromBundle` parameter adds the `main.css` file separately from the global bundle. You can remove that parameter to include it in the `global.css` bundle file.

Similar to the style bundle, the `index.html` file contains an `ABP:Scripts` part, as shown in the following code block:

```
<!--ABP:Scripts-->
<script src="global.js?_v=637680281013693676"></script>
<!--/ABP:Scripts-->
```

Again, this code part is created (and updated) by the ABP CLI with the `abp bundle` command. If you want to include files, you can do it inside the `AddScripts` method of your bundle contributor class. The paths of the files are considered relative to the `wwwroot` folder.

Summary

This chapter was a quick introduction to the ABP Framework Blazor UI to understand its architecture and the services you will frequently use in your applications.

Authentication is one of the most challenging aspects of an application, and ABP provides an industry-standard solution that you can directly use in your applications.

We've learned about the services to get the current user's identity information, check the user's permissions, and localize the user interface. We also explored the services to show message boxes, notifications, and alerts to the user.

ABP's dynamic C# client proxy system makes it super-easy to consume server-side HTTP APIs. Finally, you've learned how to use the global bundling system to handle bundling and minification in your Blazor applications.

I intentionally didn't cover two topics in this chapter. The first one is Blazor itself. It is a very detailed topic to cover in a single chapter of a book. I advise you to read Microsoft's documentation (`https://docs.microsoft.com/en-us/aspnet/core/blazor`) or purchase a dedicated book if you are new to the Blazor framework. Check out the book *Web Development with Blazor*, by *Jimmy Engström*, from *Packt Publishing*.

The second topic I haven't covered in this chapter is complex UI components, such as data tables, modals, and tabs. They are so specific to the UI kit you are using. ABP comes with the Blazorise library, and you can refer to its documentation to learn its components: `https://blazorise.com/docs`. I also suggest going through ABP Framework's Blazor UI tutorial to understand the fundamental development model with the most-used components, data tables, and modals: `https://docs.abp.io/en/abp/latest/Getting-Started`.

In the next chapter, we will focus on building HTTP APIs and consuming them in client applications. We will also look at using the SignalR library for real-time communication between a client and a server.

14
Building HTTP APIs and Real-Time Services

Exposing an HTTP API endpoint is a fairly common way of allowing client applications to consume your application functionalities. Building HTTP APIs makes your application open to any client since almost all devices that connect to a network already implement the HTTP protocol.

In this chapter, you will learn about options to create HTTP APIs for your solutions. You will also see how ABP makes it easy to consume your HTTP APIs from client applications by using ABP's dynamic and generated client-side proxies. Finally, we will explain how you can use Microsoft's **SignalR** library in ABP applications to implement real-time server-client communication. Here is a list of topics covered in this chapter:

- Building HTTP APIs
- Consuming HTTP APIs
- Using SignalR with ABP Framework

Technical requirements

If you want to follow the examples in this chapter, you need to have an IDE/editor that supports ASP.NET Core development. We will use the ABP CLI at some points, so you need to install the ABP CLI, as explained in *Chapter 2, Getting Started with ABP Framework.*

You can download the example application from the following GitHub repository: `https://github.com/PacktPublishing/Mastering-ABP-Framework`. It contains some of the examples given in this chapter.

Building HTTP APIs

In this section, we will begin with ASP.NET Core's standard approach for creating HTTP APIs. Then we will see how ABP can automatically convert standard application services to HTTP API endpoints. But first, let's see how we can create API-only solutions with ABP Framework.

Creating an HTTP API project

When you create a new application or module with ABP Framework's startup solution templates, it already contains HTTP APIs for all the functionality provided by the application. However, it is also possible to create an HTTP API endpoint without an application UI if you want to.

You can use the `-u none` parameter when you create a new solution using ABP Framework, as shown in the following example:

```
abp new ApiDemo -u none
```

`ApiDemo` is our solution name here. In this way, we have a solution with an HTTP API endpoint but without a UI. The following figure shows the solution opened in Visual Studio:

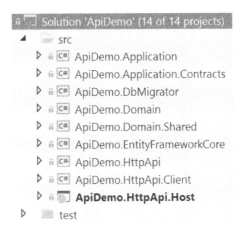

Figure 14.1 – An HTTP API solution created by the ABP CLI

We should first run the `ApiDemo.DbMigrator` application to create the database, so the HTTP API properly works. To do that, right-click the `ApiDemo.DbMigrator` project, click the **Set as Startup Project** action, then hit *Ctrl + F5* to run it. If you are not using Visual Studio, open a command-line terminal in the root directory of the `ApiDemo.DbMigrator` project and execute the `dotnet run` command.

Now, you can run the `ApiDemo.HttpApi.Host` project to start the HTTP API application. The HTTP API application shows Swagger UI by default, as shown in the following figure:

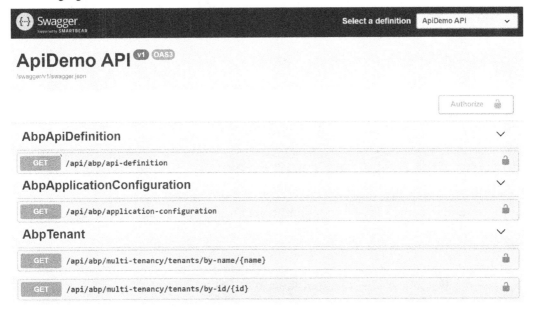

Figure 14.2 – Swagger UI

Swagger UI is a very useful tool to explore and test our HTTP API endpoints. We can use the **Authorize** button to log in to the application (the default username is admin, and the default password is 1q2w3E*), so we can also test the APIs that require authorization.

For example, we can use the /api/identity/roles endpoint to get a list of roles defined in the system. This endpoint requires authorization, so log in first with the **Authorize** button. After login, find the /api/identity/roles endpoint under the **Role** group, click to expand it, click the **Try it out** button, then the **Execute** button to call the endpoint. When you call it, the server returns a JSON value as shown in the following example:

```json
{
  "totalCount": 1,
  "items": [
    {
      "name": "admin",
      "isDefault": false,
      "isStatic": true,
      "isPublic": true,
      "concurrencyStamp":
          "1f23ae3a-85d8-4656-b094-00e605e28e4e",
      "id": "92692d73-4acb-ca9f-4838-39ff4cdf25e4",
      "extraProperties": {}
    }
  ]
}
```

So, we've learned how to create and launch an HTTP API solution with ABP Framework. Now, let's see how we can add new APIs using ASP.NET Core's standard controllers.

Creating ASP.NET Core controllers

ASP.NET Core's controllers provide a convenient infrastructure to create HTTP APIs. The following example exposes HTTP endpoints to get the product list and update a product:

```
[ApiController]
[Route("products")]
public class ProductController : ControllerBase
{
    [HttpGet]
```

```
    public async Task<ProductDto> GetListAsync()
    {
        // TODO: implement
    }

    [HttpPut]
    [Route("{id}")]
    public async Task UpdateAsync(Guid id, ProductUpdateDto
                                    input)
    {
        // TODO: implement
    }
}
```

The ProductController class is inherited from the ControllerBase class. It is suggested to inherit your API controller classes from the ControllerBase class instead of the Controller class since it contains some view-related functionality that is unnecessary for API Controllers. Alternatively, you can inherit your API controller classes from the AbpControllerBase class, which provides some common ABP services as pre-injected properties for you.

Adding the [ApiController] attribute on top of the controller class enables ASP.NET Core's default API-specific behaviors (such as automatic HTTP 400 responses and attribute routing requirement), so it is also suggested.

In this example, the [Route] attribute defines the URL of the APIs while the HttpGet and HttpPut attributes determine the HTTP method associated with the API endpoints.

ABP is 100% compatible with ASP.NET Core's standard structures, so you can refer to Microsoft's documentation to learn all the details of creating API Controllers: https://docs.microsoft.com/en-us/aspnet/core/web-api.

When you implement layering in your solution, you generally find yourself creating controller classes, which wrap your application services. For example, assume that you have IProductAppService, which already implements the product-related use cases, and you want to expose its methods as HTTP API endpoints. The following example defines a controller that redirects all requests to the underlying application service:

```
[ApiController]
[Route("products")]
public class ProductController : ControllerBase
```

```
{
    private readonly IProductAppService _productAppService;

    public ProductController(
        IProductAppService productAppService)
    {
        _productAppService = productAppService;
    }


    [HttpGet]
    public async Task<ProductDto> GetListAsync()
    {
        return await _productAppService.GetListAsync();
    }


    [HttpPut]
    [Route("{id}")]
    public async Task UpdateAsync(Guid id,
                                    ProductUpdateDto input)
    {
        await _productAppService.UpdateAsync(id, input);
    }
}
```

If we didn't use ABP Framework, we'd need to write such controllers to be able to define the route, HTTP method, and other HTTP-related details for the endpoint. However, ABP Framework can automatically expose your application services as HTTP API endpoints, as explained in the next section.

Understanding the Auto API Controllers

ABP's Auto API Controller system converts your application services to API controllers by convention. To enable Auto API Controllers, we should configure `AbpAspNetCoreMvcOptions` as shown in the following code block:

```
Configure<AbpAspNetCoreMvcOptions>(options =>
{
    options.ConventionalControllers.Create(
        typeof(ApiDemoApplicationModule).Assembly);
});
```

That configuration code is located in the UI or HTTP API layer of the solution (the `ApiDemoHttpApiHostModule` class of the **ApiDemo.HttpApi.Host** project for this example). The `options.ConventionalControllers.Create` method takes an `Assembly` object, finds all the application service classes inside that `Assembly`, and exposes them as controllers using pre-defined conventions. When you create a new ABP solution from the startup template, you already have that configuration inside your solution, so you don't need to configure it yourself.

Assume that we have defined an application service as in the following example:

```
public class ProductAppService
    : ApiDemoAppService, IProductAppService
{
    public Task<ProductDto> GetListAsync()
    {
        // TODO: implement
    }

    public Task UpdateAsync(Guid id,
                            ProductUpdateDto input)
    {
        // TODO: implement
    }
}
```

Remember that the `ProductAppService` class is defined in the `ApiDemo.Application` project and the `IProductAppService` interface is defined in the `ApiDemo.Application.Contracts` project. We can run the application without any additional configuration to see the new HTTP API endpoints on Swagger UI:

Figure 14.3 – Auto API Controller on Swagger UI

ABP Framework configured ASP.NET Core so `ProductAppService` becomes a controller. ABP Framework determines the HTTP method automatically by the name of the related C# method. For example, methods starting with the `Get` prefix are considered HTTP GET methods. Routes are also automatically determined by convention. You can refer to the ABP documentation to learn about all the conventions and customization options for the HTTP method and route determination: `https://docs.abp.io/en/abp/latest/API/Auto-API-Controllers`.

> **When to Define Controllers Manually**
>
> When you use ABP Framework, you generally don't need to define the API controllers manually. However, you can still write the controllers in a standard way if you want to do so. An advantage of writing manual controllers is that you can fully use HTTP layer capabilities to define and shape your APIs.

ABP Framework converts all the application services to API controllers in the configured assembly. If you want to disable it for a specific application service, you can use the `[RemoteService]` attribute with the `false` parameter as shown in the following example:

```
[RemoteService(false)]
public class ProductAppService
        : ApiDemoAppService, IProductAppService
{ /* ... */ }
```

ABP Framework also enables ASP.NET Core's API explorer feature for the application service. In this way, your API endpoints become discoverable and are shown on Swagger UI. If you want to expose the HTTP endpoint but disable the API explorer, you can set the `IsMetadataEnabled` parameter of the `[RemoteService]` attribute to `false`, for example, `[RemoteService(IsMetadataEnabled = false)]`.

As we've learned in this section, ABP can automate exposing your application services to remote clients, while you can still use your existing skills to create standard ASP.NET Core controllers whenever you need them. In the next section, we will explore the ways you consume your HTTP APIs from client applications.

Consuming HTTP APIs

Consuming your HTTP APIs from your client applications typically requires a lot of common and repetitive logic to apply. You deal with authorization, object serialization, exception handling, and more in every HTTP request to the server. ABP Framework can completely automate that process via dynamic and generated (static) client-side proxies.

We've already covered the practical usage of ABP's client-side proxy system in the *Consuming HTTP APIs* section of *Chapter 12, Working with MVC/Razor Pages*, and in the *Consuming HTTP APIs* section of *Chapter 13, Working with the Blazor WebAssembly UI*. So, I won't repeat it here but will bring it all together and fill in the missing points.

Let's start with dynamic client proxies.

Using ABP's dynamic client-side proxies

The dynamic proxy system allows us to consume server-side HTTP APIs with a simple configuration. The *dynamic* name states that the proxy code is generated dynamically at runtime.

ABP's dynamic client proxy system supports two types of client applications: .NET and JavaScript.

Using dynamic .NET client proxies

ABP's startup solution separates the application layer into two projects. The project that ends with `Application.Contracts` contains the interfaces and **data transfer objects (DTOs)** for our application services, while the project ending with `Application` contains the implementation of these interfaces. The following figure shows the `IProductAppService` interface and the `ProductAppService` class inside the `Application.Contracts` and the `Application` projects in an example solution:

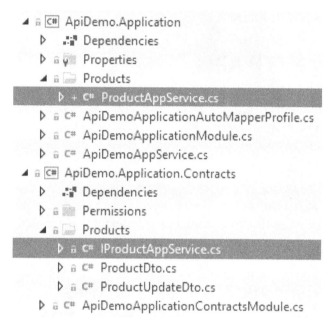

Figure 14.4 – The application layer, separated into two projects

Separating the contracts from the implementation has an advantage: we can reuse the `ApiDemo.Application.Contracts` project from a .NET client application without making the client application reference the implementations of the application services.

A .NET client application can reference the `ApiDemo.Application.Contracts` project and configure ABP's dynamic .NET client proxy system to be able to consume the HTTP APIs, just like consuming a local service. The following example shows that configuration, which is done in a client application (that configuration exists in the `ApiDemoHttpApiClientModule` class of the `ApiDemo.HttpApi.Client` project):

```
public override void ConfigureServices(
    ServiceConfigurationContext context)
{
```

```
        context.Services.AddHttpClientProxies(
            typeof(ApiDemoApplicationContractsModule).Assembly
        );
}
```

The AddHttpClientProxies method takes an Assembly and creates dynamic proxies for all the application service interfaces in that Assembly. Here, we pass the assembly of the ApiDemo.Application.Contracts project by using the module class inside it. When you create a new ABP solution, you will see that configuration in the HttpApi.Client project. So, any .NET client application that references the HttpApi.Client project can directly consume your HTTP APIs without any configuration.

With such a simple single configuration, we can inject any application service interface into the ApiDemo.Application.Contracts project and use it as we use a local service. See the *Consuming HTTP APIs* section of *Chapter 13, Working with the Blazor WebAssembly UI*, for example usage in a Blazor WebAssembly client application.

Once we configure and use the dynamic .NET client proxies, ABP does all the heavy logic and performs an HTTP request to the server for us. Surely, ABP should know the server's root URL to make the request. We can define it in the appsettings.json file of the client application as shown in the following example (there is an example of how to do that inside the ApiDemo.HttpApi.Client.ConsoleTestApp project):

```
{
  "RemoteServices": {
    "Default": {
      "BaseUrl": "http://localhost:53929/"
    }
  }
}
```

As you can understand from this example, we can actually define multiple server endpoints. In this way, a client application can consume APIs from more than one server. The Default configuration one is used by default. You can add a second remote service configuration as shown in the following example:

```
{
  "RemoteServices": {
    "Default": {
      "BaseUrl": "http://localhost:53929/"
```

```
    },
    "BookStore": {
      "BaseUrl": "http://localhost:48392/"
    }
  }
}
```

Then you should specify the `remoteServiceConfigurationName` parameter to the `AddHttpClientProxies` method to map the configurations:

```
context.Services.AddHttpClientProxies(
    typeof(BookStoreApplicationContractsModule).Assembly,
    remoteServiceConfigurationName: "BookStore"
);
```

You can add retry logic on failure to the dynamic client proxies. Please refer to the documentation for more configuration options: `https://docs.abp.io/en/abp/latest/API/Dynamic-CSharp-API-Clients`.

ABP Framework provides a special API endpoint from the server application that exposes the API definition to the clients. This endpoint contains the mapping between the application service contracts and the HTTP API endpoints of the application. The URL of that endpoint is `/api/abp/api-definition` on the server. Client applications first read that API definition endpoint to learn how to make HTTP calls to the server.

As you've seen, ABP makes it extremely easy to consume HTTP APIs from .NET clients. In the next section, we will look at consuming HTTP APIs in a JavaScript client.

Using dynamic JavaScript client proxies

Like the .NET dynamic client proxies, ABP dynamically creates proxies to consume your HTTP API endpoints from JavaScript applications. ABP Framework provides a special endpoint that returns a JavaScript file that contains proxy functions for all of your HTTP API endpoints. The URL of the endpoint is `/Abp/ServiceProxyScript`. This URL is already added to the application layout by the current theme, so you can directly consume the HTTP APIs.

The following code block is a part of the service proxy script endpoint that contains the proxy functions for the `ProductAppService` class that we previously created in the *Understanding the Auto API Controllers* section of this chapter:

```
apiDemo.products.product.getList = function(ajaxParams) {
  return abp.ajax($.extend(true, {
    url: abp.appPath + 'api/app/product',
    type: 'GET'
  }, ajaxParams));
};


apiDemo.products.product.update =
    function(id, input, ajaxParams) {
  return abp.ajax($.extend(true, {
    url: abp.appPath + 'api/app/product/' + id + '',
    type: 'PUT',
    dataType: null,
    data: JSON.stringify(input)
  }, ajaxParams));
};
```

As you can see in that example, ABP Framework has created two JavaScript functions for the `ProductAppService` class's methods. For example, we can call the `getList` function to get a list of the products, as shown in the following example:

```
apiDemo.products.product.getList()
  .then(function(result){
    // TODO: Process the result...
  });
```

It's that easy! Authorization, validation, exception handling, **CSRF (Cross-Site Request Forgery)**, and the other details are handled by ABP Framework. The result value will be the product list (array) returned by the server. You can see the *Using dynamic client proxies* section of *Chapter 12, Working with MVC/Razor Pages*, for more examples and information about this topic.

In the next section, we will explore an alternative way to use dynamic client proxies.

Using ABP's static (generated) client proxies

The dynamic proxy system completely automates proxy generation to consume HTTP endpoints from the client applications. It generates code at runtime based on dynamically obtained endpoint configuration.

On the other hand, the static client proxy system (that comes with ABP v5.0) doesn't require obtaining the API definitions at runtime since it generates the client proxy code at development time. The disadvantage of the static proxy system is that you need to re-generate the client proxy code whenever a server API changes. However, static proxies are slightly faster than dynamic proxies since the code generation is done at development time and no runtime information is required.

In some scenarios, such as when your client consumes HTTP APIs of multiple microservices behind an API gateway, the dynamic client proxy system can't directly work because the API gateway cannot combine and return the API definitions of all microservices from a single endpoint. In such cases, using static client proxies that were generated at development time can save us.

In any case if you want to use the static client proxies, you use the ABP CLI to generate the client code. The following section shows how to use the ABP CLI to generate static C# client proxy code.

Generating static C# client proxies

In order to create static proxies, the client application/project should have a reference to the application service interfaces defined by the server because the client proxies implement the same interfaces and are used just like the dynamic proxies. So, in practice, the client application should reference the `Application.Contracts` project of the target application.

The server application should be running when we use the ABP CLI to generate the proxy classes since the ABP CLI gets the API definition from the server. Once the server is up and running, use the `generate-proxy` command in the root folder of the client application/project, as shown in the following example:

```
abp generate-proxy -t csharp -u https://localhost:44367
```

`https://localhost:44367` is the server application's URL here. The `-t` parameter specifies the client language, which is `csharp` for this example.

The following figure shows the newly added project files after running the `generate-proxy` command:

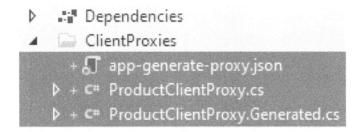

Figure 14.5 – Generated client proxy files

First of all, the ABP CLI adds the `app-generate-proxy.json` file, which contains the API definition obtained from the `https://localhost:44367/api/abp/api-definition` endpoint for this example. ABP Framework then uses this file to get information about the API endpoint and make proper HTTP calls.

The `ProductClientProxy.Generated.cs` file contains the proxy class, which implements the `IProductAppService` interface for this example. In this way, we can inject the `IProductAppService` interface into any class and use it just like a local service. ABP performs the necessary HTTP API calls for us.

`ProductClientProxy.cs` is a partial class to add your additional methods and customize the class. The `ProductClientProxy.Generated.cs` file is re-generated whenever you execute the `generate-proxy` command, so your changes are overwritten if you edit that class. However, the `ProductClientProxy.cs` file can be safely edited since ABP won't touch it again. It is left for you to customize the class.

In the next section, we will generate JavaScript proxies to consume HTTP APIs from a browser application.

Generating static JavaScript client proxies

The ABP CLI can generate HTTP API client proxies for JavaScript clients, just like .NET clients. We can specify the `-t` parameter as `js` for JavaScript code generation:

```
abp generate-proxy -t js -u https://localhost:44367
```

The JavaScript client proxy system works on jQuery and is compatible with ABP's MVC/Razor Pages UI. We've already seen the usage of JavaScript client code generation in the *Using static client proxies* section of *Chapter 12, Working with MVC/Razor Pages*. Please refer to that chapter to remember its usage.

Generating static Angular client proxies

While not covered in this book, ABP has a first-class Angular UI integration option. The ABP CLI's generate-proxy command also works natively with Angular UI. You can specify the -t parameter as ng to generate TypeScript proxy code for Angular:

```
abp generate-proxy -t ng -u https://localhost:44367
```

The ABP CLI creates services and DTO classes on the Angular side, so you can directly inject proxies and consume the HTTP APIs without dealing with low-level HTTP details. Please refer to the ABP documentation to learn more about Angular client proxies: https://docs.abp.io/en/abp/latest/UI/Angular/Service-Proxies.

Generating proxies for other client types

ABP provides client proxy generation for the client types it supports out of the box. It is suggested to use the ABP CLI's code generation for the supported client types. However, you may use another type of language, framework, or library on the client side and may want to generate the client proxies instead of manually writing them. In this case, you can use another tool that supports your platform since the ABP startup solution is compatible with Swagger/OpenAPI specifications. There are many tools around that can read the Swagger/OpenAPI specification and generate client-side proxy code for you. For example, the NSwag tool can generate client proxies for many different languages.

We've learned how to consume server-side HTTP APIs from our client applications with ABP Framework. In the next section, we will learn how to establish a real-time communication channel with the server using Microsoft's SignalR library.

Using SignalR with ABP Framework

Building REST-style HTTP APIs is good to consume server-side functionalities from client applications. However, it is limited – only the client application can call server APIs, and the server cannot normally start an operation on the client. WebSocket technology makes it possible to establish a two-way communication channel between the browser and the server to send messages to each other independently. So, with WebSocket, the server can notify the browser, send data, and trigger an action on the application.

SignalR is a library by Microsoft that runs on WebSocket technology and simplifies the communication between the server and the client by abstracting WebSocket details. You can directly call the methods defined on the client from the server and vice versa.

ABP Framework does not add much value to SignalR since it is already easy to use. However, it provides a simple integration package that automates some common tasks for you. In the next two sections, we will see how to install and configure SignalR in your solutions. Let's start with ABP's server-side SignalR integration package.

Using the ABP SignalR integration package

`Volo.Abp.AspNetCore.SignalR` is the NuGet package that adds the SignalR library to your server-side ABP application. You can install it using the ABP CLI. Open a command-line terminal in the root directory of the project where you want to add a server-side SignalR endpoint and execute the following command:

```
abp add-package Volo.Abp.AspNetCore.SignalR
```

The ABP CLI will install the NuGet package and add ABP module dependency for you. It also adds SignalR to dependency injection and configures the hub endpoint. So, you don't need an additional configuration after the installation. The next sections explain how to create SignalR hubs and do additional configuration when you need to.

Creating hubs

SignalR hubs are used to create a high-level pipeline to handle client-server communication. You should define at least one hub to use SignalR. Creating a hub is pretty easy; just define a new class derived from the `Hub` base class:

```
public class MessagingHub : Hub
{

}
```

ABP automatically registers the hub to the dependency injection system and configures the endpoint mapping. The URL of this example hub will be `/signalr-hubs/messaging`. The hub URL starts with `/signalr-hubs/` and continues with the hub class name converted to *kebab-case* without the `Hub` suffix. You can use the `[HubRoute]` attribute on top of the hub class to specify a different URL, as shown in the following example:

```
[HubRoute("/the-messaging-hub")]
public class MessagingHub : Hub
{
    //...
}
```

As an alternative to the Hub class, you can inherit your hub from the AbpHub class. The AbpHub class provides some common services (such as ICurrentUser, ILogger, and IAuthorizationService) pre-injected as base properties, so you don't need to inject them manually.

Configuring hubs

ABP automatically maps your hubs and does the basic configuration. If you want to customize the hub configuration, you can do it in the ConfigureServices method of your module class, as shown in the following example:

```
Configure<AbpSignalROptions>(options =>
{
    options.Hubs.AddOrUpdate(
        typeof(MessagingHub),
        config => //Additional configuration
        {
            config.RoutePattern = "/the-messaging-hub";
            config.ConfigureActions.Add(hubOptions =>
            {
                hubOptions.LongPolling.PollTimeout =
                    TimeSpan.FromSeconds(30);
            });
        }
    );
});
```

This example configures MessagingHub, sets a custom route, and changes the LongPolling options.

Using the ABP SignalR integration package doesn't add much value but simplifies integrating and configuring the SignalR library on the server side of our ABP applications. In the next section, we will see ways of connecting to a SignalR hub from a client application.

Configuring SignalR clients

Connecting to a SignalR hub from a client application depends on your client type. In this section, I will explain how to install the SignalR client library to an ABP application with the ASP.NET Core MVC UI. Please refer to Microsoft's documentation for other client types, such as TypeScript or .NET clients: `https://docs.microsoft.com/en-us/aspnet/core/signalr`.

To install SignalR in an ABP application with the ASP.NET Core MVC UI, first, add the `@abp/signalr` NPM package to your web project with the following command:

```
npm install @abp/signalr
```

This command will install the package and update the `package.json` file in the web project. You should then run the ABP CLI's `install-libs` command to copy SignalR's JavaScript file under your project's `wwwroot/libs` folder.

After the installation, you can use SignalR in your pages by importing SignalR's JavaScript file. You can use ABP's `abp-script` tag helper with the pre-defined bundle contributor for SignalR, as in the following example:

```
@using Volo.Abp.AspNetCore.Mvc.UI.Packages.SignalR
@section scripts {
    <abp-script type=
        "typeof(SignalRBrowserScriptContributor)" />
}
```

Using `SignalRBrowserScriptContributor` is the suggested approach since it always adds the script file from the right path with the right version, so you don't need to change it when you upgrade the SignalR package.

Using SignalR with ABP Framework is no different from using it in a regular ASP.NET Core application. So, please refer to Microsoft's documentation if you are new to SignalR: `https://docs.microsoft.com/en-us/aspnet/core/signalr`. You can also find a fully working example in ABP's official samples: `https://docs.abp.io/en/abp/latest/Samples/Index`.

Summary

In this chapter, you've learned different methods for server-client communication with ABP Framework and ASP.NET Core. ABP Framework automates that communication wherever possible.

We started by creating REST-style HTTP APIs with the standard ASP.NET Core controllers and learned how ABP can automatically create such controllers using the application services.

We then explored various ways of consuming the HTTP APIs from different clients. It becomes very simple to call server-side APIs from a client application when you use the dynamic or static client proxies of ABP Framework. While you can always go your own way, using the fully integrated client proxies is the best way to consume your own HTTP APIs.

Finally, we saw how we can install SignalR in your ABP applications using the pre-built integration packages. SignalR, with the WebSocket technology, makes it possible to establish a two-way communication channel between the server and the client, so the server can also send messages to the client whenever it needs to.

In the next chapter, we will learn about one of the most powerful structures of ABP Framework: modular application development.

Part 5: Miscellaneous

This part contains various topics, including the infrastructure provided by ABP Framework. You will learn how ABP handles modularity and how you can create your own modules by referencing an example case. You will explore and understand ABP's multi-tenancy infrastructure, which is used to create SaaS applications. In the final chapter, you will learn about creating unit and integration tests with ABP Framework.

In this part, we include the following chapters:

- *Chapter 15, Working with Modularity*
- *Chapter 16, Implementing Multi-Tenancy*
- *Chapter 17, Building Automated Tests*

15
Working with Modularity

Let me state at the beginning of this chapter – modular application development is hard work! We want to split a large system into smaller modules and isolate them from each other. However, then we will have difficulties when integrating these modules and making them communicate with each other.

One of the fundamental design goals of ABP Framework is modularity. It provides the necessary infrastructure to build truly modular systems.

This chapter will start with what modularity means and the levels of modularity in the .NET platform. In the largest part of the chapter, we will explore the Payment module that I've built for the EventHub reference solution. We will learn how the module is structured, the key points of application module development, and how to install the module into the main application.

This chapter consists of the following main topics:

- Understanding modularity
- Building the Payment module
- Installing the Payment module into EventHub

Technical requirements

You can clone or download the source code of the EventHub project from GitHub: `https://github.com/volosoft/eventhub`.

If you want to follow the examples in this chapter, you need to have an IDE/editor that supports ASP.NET Core development.

Finally, if you want to create modules with the ABP CLI, you should install it on your computer, as explained in the *Installing the ABP CLI* section of *Chapter 2*, *Getting Started with ABP Framework*.

Understanding modularity

The term **module** is one of the most overused and overloaded concepts in the software industry. In this section, I want to explain what I mean by modularity in .NET and ABP Framework.

Modularity is a software design technique to separate a large solution's code base into smaller, isolated modules that can then be developed independently. There are two main reasons behind modular application development:

- **Reducing complexity**: Splitting a large code base into a smaller and isolated set of modules makes it easy to develop and maintain the solution.

- **Reusability**: Building a module and reusing it across multiple applications reduces code duplication and saves time.

In the next sections, I will discuss two different modularity levels from technical and design perspectives: class libraries (NuGet packages) and application modules. Let's begin with class libraries.

Class libraries and NuGet packages

Most programming languages and frameworks have the concept of a module. In general, a module is a set of code files (classes and other resources) developed and shipped (deployed) together.

A module provides some components and services for a larger application. A module may depend on other modules and can use the components and services provided by the dependent modules.

In .NET, an assembly is a typical way to create a module. We can create a **class library project** and then use it within other libraries and applications. We can create NuGet packages for the class libraries and publish them on NuGet.org publicly. If the library is not public, we can host a private NuGet server in our own company. A NuGet package system makes it extremely easy to add a library to a project. There are thousands of packages already published on NuGet.org.

ABP Framework itself is designed to be modular. It consists of hundreds of NuGet packages; each provides different infrastructure features for your applications. Some example packages are Volo.Abp.Validation, Volo.Abp.Authorization, Volo.Abp.Caching, Volo.Abp.EntityFrameworkCore, Volo.Abp.BlobStoring, Volo.Abp.Auditing, and Volo.Abp.Emailing. You can use any package you need in your application.

You can refer to the *Understanding modularity* section of *Chapter 5*, *Exploring the ASP. NET Core and ABP Infrastructure* to learn about package-based ABP modules. The next section will discuss application modules, which typically consist of multiple packages (class library projects).

Application modules

We can think of an application module as a vertical slice of an application. An application module has the following attributes:

- Defines some business objects (for example, aggregates, entities, and value objects)
- Implements business logic for the business objects it defines
- Provides database integration and mappings for the business objects
- Contains application services, data transfer objects, and HTTP APIs (controllers)
- Can have user interface components and pages related to the functionality it provides
- May need to add new items to the application menu, layout, or toolbars on the UI
- Publishes and consumes distributed events
- May have more features and other details you expect from a regular application

There are several *isolation levels* for an application module based on your requirements and goals. Four common examples are listed here:

- **Tightly coupled modules**: A module can be a part of a large monolith application with a single database. You can use that module's entities and services in other modules and perform database queries by joining the tables of that module. In this way, your modules become tightly coupled to each other.

- **Bounded contexts**: A module can be a part of a large monolith application, but it hides its internal domain objects and database tables from other modules. Other modules can only use its integration services and subscribe to the events published by that module. They can't use the database tables of the module in SQL queries. The module may even use a different kind of DBMS for its specific requirements. That is the bounded context pattern in domain-driven design. Such a module is a good candidate to convert to a microservice if you want to convert your monolith application to a microservice solution in the future.

- **Generic modules**: Generic modules are designed to be application-independent. They can be integrated into different kinds of applications. The application that uses the generic module can have some functionalities depending on that module, and it may need some integration code. A generic module may provide some options and customization points but doesn't make assumptions about the final application. Infrastructure modules, such as identity management and multi-language modules, fall into this category. Also, the Payment module, which is explained in the *Building the Payment module* section, is a generic module.

- **Plugin modules**: A plugin module is a completely isolated and reusable application module. Other modules have no direct dependency on that module. You can easily add this module to or remove it from an existing solution without affecting the other modules and your application. If other modules need to use that module, they use some standard abstractions provided in a shared library. In this case, the module implements the abstractions and can be replaced by another module that implements the same abstractions. Even if the other modules use that module somehow, they can continue to work as expected when removing that module. That means the module should be optional and removable for the application.

One of ABP Framework's main goals is to provide a convenient infrastructure to develop any kind of application module. It provides the necessary infrastructure details to build a truly modular system. It also provides some pre-built application modules you can directly reuse in your applications. Some examples are as follows:

- An **account module** provides authentication features, such as login, register, forgot password, and social login integrations.

- An **identity module** manages users, roles, and their permissions in your system.

- A **tenant management** module allows you to create and manage tenants in an SaaS/multi-tenant system.

- A **CMS Kit** module can be used to add fundamental **Content Management System (CMS)** features into your application, such as pages, tags, comments, and blogs.

Account, identity, tenant management, and some other modules are pre-installed (as NuGet packages) when creating a new ABP solution.

All pre-built modules are designed to be extensible and customizable. However, if you need to fully change a module based on your requirements, it is always possible to download the module's source code and include it in your solution.

In the next section, we will see how to build a new application module with its own entities, services, and pages.

Building the Payment module

You can already investigate the source code of the pre-built ABP modules to see how they are built and used in your application. I suggest it because you can see different implementation details of modular development. However, this section will explore the Payment module, which has been created as a simple yet real-world example for this book. It is used by the EventHub solution to receive payment when an organization wants to upgrade to a premium account. It is not possible to show the step-by-step development of that module in a book. We will investigate the fundamental points so that you can understand the module structure and build your own modules. Let's start with creating a new application module.

Creating a new application module

The ABP CLI's new command provides an option to create a new solution to build a reusable application module. See the following example:

```
abp new Payment -t module
```

I've specified to use the module template (-t module) with the module name as Payment. If you open the solution, you will see the solution structure, as shown in the following figure:

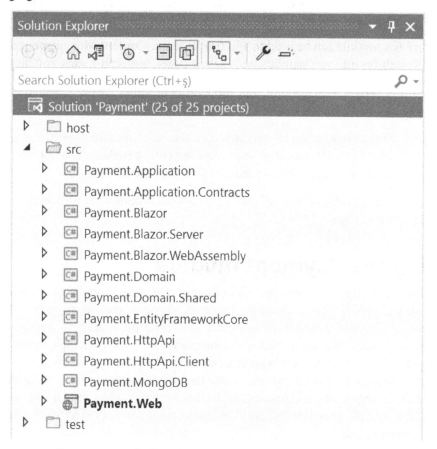

Figure 15.1 – A fresh application module created by the ABP CLI

The module startup template has too many projects because it supports multiple UI and database options and contains some test/demo projects. Let's eliminate some of the projects:

- The projects in the host folder are some demo applications to run the module in different architecture options. These projects are not parts of the module and are just for manual testing. We will install this module into the EventHub solution and test it there, so I deleted all host projects.

- I deleted the Blazor.* projects since my main UI will be MVC/Razor Pages.

- I deleted the MongoDB-related projects since I only want to support EF Core with my module.

- Finally, I deleted the angular folder (not shown in *Figure 15.1*) since I don't want an Angular UI for this module.

After the cleanup, there are 12 projects in the module solution. Four of them are for unit and integration testing, so the module consists of eight projects that will be deployed. These eight projects are class library projects, so they can't be run individually. They need to be used by an executable application, such as EventHub:

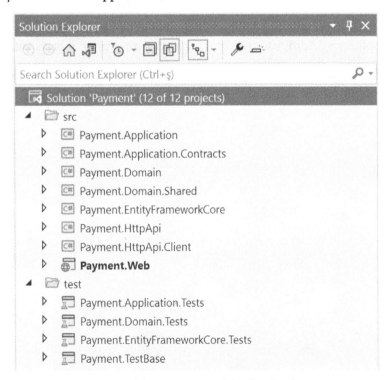

Figure 15.2 – The Payment module after the cleanup

This solution structure and layers were already explained in the *Structuring a .NET solution based on DDD* section of *Chapter 9, Understanding Domain-Driven Design*. So, I won't repeat it all here. However, we will change this structure because we want to provide multiple application layers for the Payment module.

Restructuring the Payment module solution

We will install this Payment module to the EventHub solution. Remember from *Chapter 4, Understanding the Reference Solution*, that the EventHub solution has two UI applications:

- A public website that the end users of the system use to create and attend events. This application has an MVC/Razor Pages UI.

- An admin web application that is used by the administrative users of the EventHub system. This application is a Blazor WebAssembly application.

To support the same architecture, we will provide two UI application layers for the Payment module:

- An application layer with the MVC/Razor Pages UI that is used by the EventHub public website. End users will make payments with that UI.

- An application layer with the Blazor WebAssembly UI that is used by the EventHub admin application. Administrative users will see the payment reports with that UI.

The following figure shows the final Payment solution structure after I added the admin-side layers and organized the solution folders:

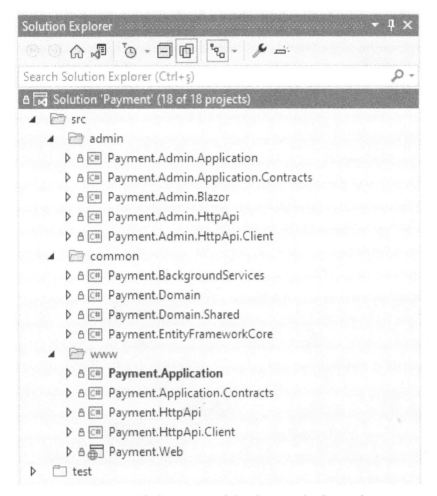

Figure 15.3 – The Payment module solution with admin-side

I added the Payment.Admin.Application, Payment.Admin.Application. Contracts, Payment.Admin.Blazor, Payment.Admin.HttpApi, and Payment.Admin.HttpApi.Client projects. I also added the Payment. BackgroundServices project to perform some periodic background workers.

The solution folders reflect the overall structure – the admin application (with Blazor UI) and www (public) application (with MVC/Razor Pages UI). The common folder is used in both applications, so we share the same domain layer and database integration code.

We've learned about the overall structure of the Payment solution. In the next section, you will learn the details of the payment process.

Understanding the payment process

The only responsibility of the Payment module is to take a payment from the user. It internally uses PayPal as the payment gateway. The Payment module is generic and can be used by any kind of application. The application that uses the Payment module should include some integration logic that starts the payment process and handles the payment result. In this section, I will explain the process based on the EventHub integration.

The EventHub application uses the Payment module to get a payment from a user to upgrade a free organization account to a premium organization account. As you can guess, premium organizations have more rights in the application.

If you are the owner of an organization and visit the organization details page, you will see an **Upgrade to Premium** button on the page, as shown in the following figure:

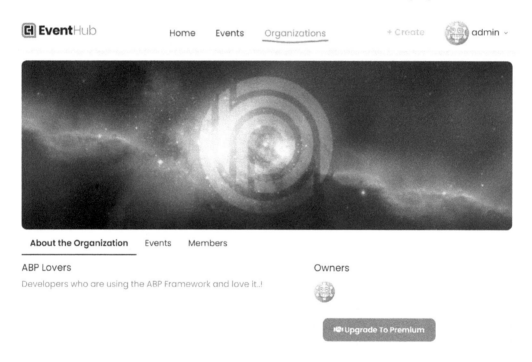

Figure 15.4 – The EventHub organization details page

When you click on the **Upgrade to Premium** button, you are redirected to the pricing page:

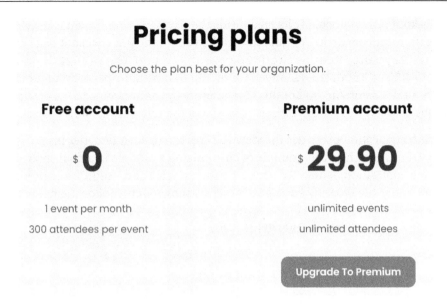

Figure 15.5 – The EventHub pricing page

Here, we can see the account types and their differences. When we click the **Upgrade to Premium** button here, we are redirected to the pre-checkout page, which is defined by the Payment module:

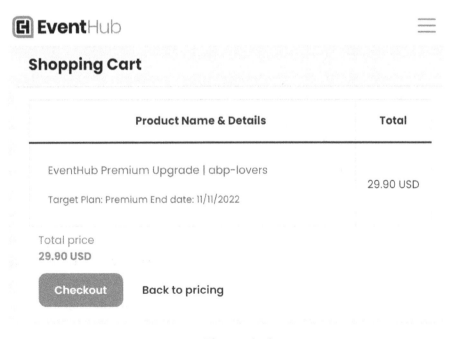

Figure 15.6 – The pre-checkout page

The pre-checkout page is normally located inside the Payment module and is developed to be application-independent. We can redirect the user to the pre-checkout page with a URL such as /Payment/PreCheckout?paymentRequestId=3a002186-cb04-eb46-7310-251e45fc6aed. However, we should first obtain a payment request ID using the CreateAsync method of the IPaymentRequestAppService service. This is done in the Pages/Pricing.cshtml.cs file of the EventHub.Web project.

The EventHub application overrides the view (UI) part to make it fit better into EventHub's UI design. This is an example of customizing a module in a final application. The EventHub application defines PreCheckout.cshtml and PostCheckout. cshtml files under the Pages/Payment folder, as shown in the following figure:

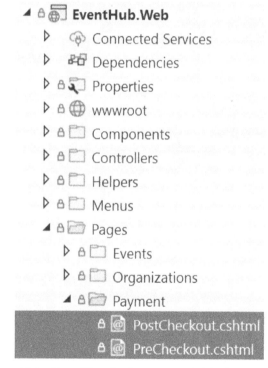

Figure 15.7 – Overriding the Payment module's checkout views

They automatically override the corresponding Payment pages (because they are located exactly within the same path defined by the Payment module). These pages have no .cshtml.cs files here because we don't want to change the behavior of the page; we just want to change the view side.

The following figure shows the main components and the flow used for the payment process:

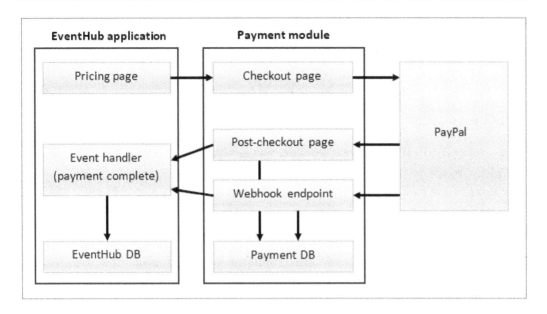

Figure 15.8 – The payment flow

When we click the **Upgrade to Premium** button on the Pricing page (in *Figure 15.5*), it redirects to the Checkout page of the Payment module. When we click the **Checkout** button (in *Figure 15.6*) on that page, we are redirected to PayPal, the payment system used and integrated by the Payment module. Once we complete the payment on PayPal, we are redirected back to the post-checkout page of the application, which shows a thank you message to the user.

When a payment process succeeds, the Payment module publishes a distributed event named `PaymentRequestCompletedEto` (defined in the `Payment.Domain.Shared` project). The EventHub application subscribes to this event (with the `PaymentRequestEventHandler` class inside the `EventHub.Domain` project). It finds the user and organization related to the completed payment, upgrades the organization, and sends an email to thank the user for upgrading the account.

There can be an error when returning from PayPal to our application in some rare cases, where we cannot know whether the payment process has succeeded. For such cases, the Payment module provides a **Webhook** endpoint that PayPal calls to inform us about the status of the payment operation. The Webhook request is handled by `PaymentRequestController` (in the `Payment.HttpApi` project). If the operation succeeds, the same `PaymentRequestCompletedEto` event is published so that the EventHub application can upgrade the organization account asynchronously.

In the next section, we will see how the Payment module provides options to configure it.

Providing configuration options

The Payment module uses PayPal, so it needs the PayPal account information that the application must configure. It follows the options pattern (see the *Implementing the options pattern* section of *Chapter 5, Exploring the ASP.NET Core and ABP Infrastructure*) and provides the `PayPalOptions` class that can be configured by the application, as shown in the following example:

```
Configure<PayPalOptions>(options =>
{
    options.ClientId = "...";
    options.Secret = "...";
});
```

We generally get the values from the configuration (the `appsettings.json` file). The Payment module can get the option values from the `Payment:PayPal` key if you've defined it, as in the following example:

```
"Payment": {
  "PayPal": {
    "ClientId": "...",
    "Secret": "...",
    "Environment": "Sandbox"
  }
}
```

This is made possible with the following code, located in the `PaymentDomainModule` class of the `Payment.Domain` project:

```
Configure<PayPalOptions>(configuration.GetSection("Payment:
                        PayPal"));
```

Getting the values from the configuration by default is a good practice.

I've introduced the main points of the Payment module's structure. Its code base is not so different from a typical ABP application. You can explore its source code to understand how it internally works. In the next section, we will see how it is installed in the EventHub application.

Installing the Payment module into EventHub

A module itself is not a runnable project. It should be installed into a larger application and work as part of it. In this section, we will see how the Payment module is installed in the EventHub solution.

Setting the project dependencies

The Payment module consists of more than 10 projects in its solution (see *Figure 15.3*). Similarly, the EventHub solution has a lot of projects with three applications – the admin-side, the public, and the account (`IdentityServer`) applications.

I want to integrate the Payment module into the EventHub solution in all layers. Typically, each layer of the EventHub solution should depend on (use) the corresponding layer of the Payment module. The following table shows all the dependencies of the EventHub projects to the Payment module projects:

EventHub project	Payment module project
`EventHub.Domain.Shared`	`Payment.Domain.Shared`
`EventHub.Domain`	`Payment.Domain`
`EventHub.EntityFrameworkCore`	`Payment.EntityFrameworkCore`
`EventHub.BackgroundServices`	`Payment.BackgroundServices`
`EventHub.Application.Contracts`	`Payment.Application.Contracts`
`EventHub.Application`	`Payment.Application`
`EventHub.HttpApi`	`Payment.HttpApi`
`EventHub.HttpApi.Client`	`Payment.HttpApi.Client`
`EventHub.Web`	`Payment.Web`
`EventHub.Admin.Application.` `Contracts`	`Payment.Admin.Application.` `Contracts`
`EventHub.Admin.Application`	`Payment.Admin.Application`
`EventHub.Admin.HttpApi`	`Payment.Admin.HttpApi`
`EventHub.Admin.HttpApi.Client`	`Payment.Admin.HttpApi.Client`
`EventHub.Admin.Web`	`Payment.Admin.Blazor`

Figure 15.9 – EventHub and Payment module project dependencies

So, we should add project references one by one. For example, we add the `Payment.Domain` project dependency to the `EventHub.Domain` project. This way, we can use Payment module entities in our application's domain layer.

Visual Studio doesn't properly support adding a local project dependency to a project from outside of a solution (I call it an external project dependency). However, we can manually add `ProjectReference` elements into the `csproj` file of the target project. So, we can add the following line into the `EventHub.Domain.csproj` file:

```
<ProjectReference Include=
    "..\..\modules\payment\src\Payment.Domain\Payment
    .Domain.csproj" />
```

When we add such an external project dependency, Visual Studio cannot automatically resolve it. We should open a command-line terminal and run the `dotnet restore` command. This command is only required when you add a new dependency or remove an existing dependency. In addition, if you want to build the EventHub solution with the Payment module, you can use the `dotnet build /graphBuild` command. While this is rarely needed, it can save the day when Visual Studio cannot resolve some types in the dependent module.

Once we add a project reference, we should also add the ABP module dependency. The following code block shows the `PaymentDomainModule` dependency of the `EventHubDomainModule` class:

```
[DependsOn(
    ...,
    typeof(PaymentDomainModule)
)]
public class EventHubDomainModule : AbpModule
{ ... }
```

We should manually set up all the project dependencies, as explained here. The next step is to configure the database tables for the Payment module.

Configuring the database integration

The Payment module needs some database tables to work properly. We can use the main EventHub database to store the Payment module's tables. With this approach, we will have a single database for the system. Alternatively, we can create a separate database for the Payment module so that we have two databases. The EventHub solution prefers the first approach since it is simpler to implement and manage. However, I will also show how we would implement the separate database approach. Let's begin with the single database approach.

Using a single database

In this section, I will show you how we create the Payment tables in the main database of the EventHub application.

The EventHub solution has an `EventHubDbContext` class inside the `EventHub.EntityFrameworkCore` project, which is the main class that maps the entities to the database tables. The Payment module defines a `ConfigurePayment` extension method that we call from the `OnModelCreating` method of our `DbContext` class to include the Payment database mapping model in our main database model (see the `EventHubDbContext` class in the `EventHub.EntityFrameworkCore` project):

```
protected override void OnModelCreating(ModelBuilder
                                        builder)
{
    base.OnModelCreating(builder);

    ...
    builder.ConfigurePayment(); // ADDED THIS LINE
    builder.ConfigureEventHub();
}
```

Here, `builder.ConfigurePayment()` is defined by the Payment module (in the `PaymentDbContextModelCreatingExtensions` class of the `Payment.EntityFrameworkCore` project). After adding this line inside the `OnModelCreating` method, we can add a new database migration to the EventHub solution using the following command in a command-line terminal (we run this command in the `root` folder of the `EventHub.EntityFrameworkCore` project):

```
dotnet ef migrations add Added_Payment_Module
```

This command creates a new migration file. Then, we can apply the new migration against the database using the following command:

```
dotnet ef database update
```

That's all. The Payment module will use the main EventHub database to store its data. This way, we will have a single database that contains all the tables of the application. In the next section, we will discuss the separate database approach.

Using a separate database

In this section, we will see how to change the EventHub solution to use a separate database for the Payment module. The EventHub solution uses PostgreSQL as the database provider. We will use Microsoft's SQL Server for the Payment module. This way, you will learn how to work with multiple database providers in a single application.

I made the changes in a separate branch and created a draft **Pull Request** (**PR**) on GitHub so that you can see all the changes:

- GitHub branch URL: `https://github.com/volosoft/eventhub/tree/payment-sepr-db`

- PR URL: `https://github.com/volosoft/eventhub/pull/74`

Here, I will point out the fundamental changes I've made. You can see the PR on GitHub for all the changes.

To begin with, I've created a second `DbContext` class, named `EventHubPaymentDbContext`, in the `EventHub.EntityFrameworkCore` project to manage the database migrations:

```
[ReplaceDbContext(typeof(IPaymentDbContext))]
[ConnectionStringName(
    PaymentDbProperties.ConnectionStringName)]
public class EventHubPaymentDbContext
    : AbpDbContext<EventHubPaymentDbContext>,
    IPaymentDbContext
{
    public DbSet<PaymentRequest> PaymentRequests { get;
                                                  set; }

    public EventHubPaymentDbContext(
        DbContextOptions<EventHubPaymentDbContext> options)
        : base(options)
    { }

    protected override void OnModelCreating(
        ModelBuilder modelBuilder)
    {
        base.OnModelCreating(modelBuilder);
```

```
            modelBuilder.ConfigurePayment();
    }
}
```

This class replaces `IPaymentDbContext` (defined by the Payment module) using the `[ReplaceDbContext]` attribute and implementing the `IPaymentDbContext` interface. It also declares the `[ConnectionStringName]` attribute to use the `Payment` connection string name instead of `Default` in the `appsettings.json` files. Finally, it calls the `modelBuilder.ConfigurePayment()` extension method of the Payment module to configure the database mappings.

The Payment module was designed to be independent of any specific **Database Management System (DBMS)**. It depends on the `Volo.Abp.EntityFrameworkCore` package, which is DBMS-agnostic. Since I wanted to use a SQL Server database, I've added the `Volo.Abp.EntityFrameworkCore.SqlServer` package dependency to the `EventHub.EntityFrameworkCore` project. I also added `AbpEntityFrameworkCoreSqlServerModule` to the `DependsOn` attribute of the `EventHubEntityFrameworkCoreModule` class since ABP requires it.

EF Core's command-line tool requires creating a `DbContext` factory class to create an instance of the related `DbContext` class when we run its commands. You can see the `EventHubPaymentDbContextFactory` class in the source code. It uses the `Payment` connection string and the `UseSqlServer` extension method to configure SQL Server as the database provider. With this change, we should add the `Payment` connection string in the `EventHub.DbMigrator` project's `appsettings.json` file:

```
"ConnectionStrings": {
  "Default": "Host=localhost;
              Database=EventHub;
              Username=root;
              Password=root;
              Port=5432",
  "Payment": "Server=(LocalDb)\\MSSQLLocalDB;
              Database=EventHubPayment;
              Trusted_Connection=True"
},
```

ABP will automatically obtain the `Payment` connection string for the new
`DbContext` class because it has the `ConnectionStringName` attribute (the value of
`PaymentDbProperties.ConnectionStringName` is `Payment`). I also need to
add the `Payment` connection string to all the `appsettings.json` files in which I have
defined the `Default` connection string.

I should register the new `EventHubPaymentDbContext` class to the dependency
injection system and configure it. To do that, I've changed the `ConfigureServices`
method of the `EventHubEntityFrameworkCoreModule` class, as follows:

```
public override void ConfigureServices(
    ServiceConfigurationContext context)
{
    context.Services.AddAbpDbContext<EventHubDbContext>(
        options =>
        {
        options.AddDefaultRepositories();
    });
    context.Services.AddAbpDbContext<
        EventHubPaymentDbContext>();
    Configure<AbpDbContextOptions>(options =>
    {
        options.UseNpgsql();
        options.Configure<EventHubPaymentDbContext>(opts =>
        {
            opts.UseSqlServer();
        });
    });
}
```

The `AddAbpDbContext<EventHubPaymentDbContext>()` call registers the new
`DbContext` class. I also added the `Configure<EventHubPaymentDbContext>(...)`
block to use SQL Server for this `DbContext` class. Other `DbContext` classes will
continue to use PostgreSQL (the `UseNpgsql()` call globally configures all
`DbContext` classes).

The `EventHub.DbMigrator` application executes database migrations for the main database. Now, we've got a second database and we want to change the `EventHub.DbMigrator` application so that it also executes database migrations for the Payment module's database. The change is simple; I've added the following code block inside the `EntityFrameworkCoreEventHubDbSchemaMigrator` class's `MigrateAsync` method:

```
await _serviceProvider
    .GetRequiredService<EventHubPaymentDbContext>()
    .Database
    .MigrateAsync();
```

This class is used by the `EventHub.DbMigrator` application while migrating the databases. So, by adding this code block, the new database is also migrated when I run the `EventHub.DbMigrator` application.

As a final change, I will remove the Payment tables from the main EventHub database and the following line from the `EventHubDbContext` class:

```
builder.ConfigurePayment();
```

Then, I can use EF Core's command-line tool to create a database migration (in the `root` folder of the `EventHub.EntityFrameworkCore` project):

```
dotnet ef migrations add "Remove_Payment_From_Main_Database"
--context EventHubDbContext
```

Differing from standard usage, I added the `--context EventHubDbContext` parameter. I specify the `DbContext` type because there are two `DbContext` classes in the `EventHub.EntityFrameworkCore` project. Once it creates the migration (which drops the Payment tables), I can apply changes to the database using the following command:

```
dotnet ef database update --context EventHubDbContext
```

Now, the main database has no Payment tables. But we haven't created the payment database yet. To do that, I can use EF Core's command-line tool to create a database migration for the payment database (in the `root` folder of the `EventHub.EntityFrameworkCore` project):

```
dotnet ef migrations add "Initial_Payment_Database" --context
EventHubPaymentDbContext --output-dir "MigrationsPayment"
```

This time, in addition to the `context` parameter that specifies the `EventHubPaymentDbContext` type, I set the `output-dir` parameter to specify the folder to create the migrations classes in it. The default folder name is `Migrations`, but this is used by the `EventHubDbContext` class, so I can't use it. I specify `MigrationsPayment` as the folder name. The following figure shows the new migration folder in the `EventHub.EntityFrameworkCore` project:

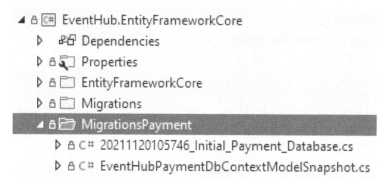

Figure 15.10 – The migration folder for the Payment module

Now, I can use the following command (in the `root` folder of the `EventHub.EntityFrameworkCore` project):

```
dotnet ef database update --context EventHubPaymentDbContexts
```

If I check the database, I can see the Payment module's table (it has a single database table):

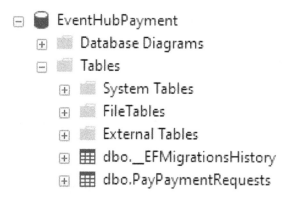

Figure 15.11 – The Payment module's table in its own database

The separate database configuration is done. Now, the Payment module will use the new SQL Server database while the rest of the application will continue to work with the main PostgreSQL database.

> **Using DbMigrator**
>
> I've used the `dotnet ef` command-line tool to update the database schema. However, we can also run the `DbMigrator` application to apply the changes to the databases. Since we've also changed the `EntityFrameworkCoreEventHubDbSchemaMigrator` class to support the second database, `DbMigrator` can migrate the database schemas for both databases.

By creating new migration `DbContext` classes, as explained here, you can set up other modules to use their own databases. I added the new `DbContext` class in the same `EventHub.EntityFrameworkCore` project; however, we could create a new project for the new `DbContext` class and manage the migrations inside it. In that case, we don't need to specify the context and the `output-dir` parameters for the EF Core commands. However, I suggest going with a single project to minimize the projects within the solution, as it already has a lot.

Summary

In this chapter, I started by explaining what modularity means and what bounded context, tightly coupled, generic, and plugin modules are. We've learned how to create a new module using the ABP CLI.

We then explored the structure of the Payment module and understood how it is integrated into the EventHub solution. We've learned the steps of manually installing the Payment module to the EventHub solution by setting up the project dependencies.

Finally, we've seen two approaches to using the payment database tables. The single database approach is simple and shares the same database between the EventHub application and the Payment module. On the other hand, the separate database approach allows us to use a dedicated database for the payment tables, making it possible to use a different DBMS for the Payment module than the main application.

I suggest checking the source code of the Payment module and the EventHub solutions to understand all the details of their structure. I also suggest you check the ABP documentation to understand modularity better and learn the best practices to build reusable, generic application modules: `https://docs.abp.io/en/abp/latest/Best-Practices/Index`.

In the next chapter, we will explore multi-tenancy, which is used to build SaaS applications.

16
Implementing Multi-Tenancy

Multi-tenancy is a common pattern to create **Software as a Service (SaaS)** solutions, where a single deployment can concurrently serve multiple customers. Multi-tenancy is one of the fundamental design principles of ABP Framework, so all other framework features are multi-tenancy compatible.

In this chapter, we will start with understanding what a multi-tenant system is and how ABP provides a multi-tenant solution to us. Then, we will continue with the ABP infrastructure to understand, build, and control the multi-tenancy in our applications. We will also learn to design specific application features and make different tenants use different application features. At the end of this chapter, you will understand the basics of multi-tenancy and will be able to build multi-tenant applications using ABP Framework.

Here is a list of the main topics covered in this chapter:

- Understanding multi-tenancy
- Working with the ABP multi-tenancy infrastructure
- Using the feature system
- When to use multi-tenancy

Technical requirements

If you want to follow the examples in this chapter, you need to have an IDE/editor that supports ASP.NET Core development.

You can download the example application from the following GitHub repository: `https://github.com/PacktPublishing/Mastering-ABP-Framework`. It contains some of the examples given in this chapter.

Understanding multi-tenancy

In this section, you will understand the SaaS and multi-tenancy concepts, the main benefits of creating a SaaS solution, and what ABP provides us as a multi-tenant-aware framework. Let's start by understanding what a SaaS system provides.

What is SaaS?

Building, deploying, and licensing software solutions with the SaaS model has become quite popular. Customers typically purchase a SaaS solution with a subscription model and use it online without requiring them to download and install it on their own servers, which is called on-premises deployment.

Building a SaaS solution has these benefits when hosting:

- You can utilize your resources at the maximum level since customers can share servers, databases, and other resources.

- It is extremely easy and typically automated to add a new customer (tenant) to the system.

- It is easier to maintain and upgrade the system compared to separate deployment for each customer.

On the other hand, using a SaaS solution benefits customers as well. They pay less for software and hosting than with an on-premises deployment. They can pay based on how much they use it. They also don't need to care about maintenance and upgrades as long as they pay the service costs.

While the SaaS solution benefits hosting, creating a SaaS solution comes with some development costs and runtime considerations.

SaaS solutions typically share the resources between customers. Some of the major shared resources are database, cache, and application servers. Data isolation, security, and performance are the main concerns that we should care about when sharing resources between customers.

In addition to shared resources, application settings, features, and permissions should be customized per customer without affecting each other.

As we now understand the benefits and challenges of building SaaS solutions, let's talk a bit about multi-tenancy.

What is multi-tenancy?

Multi-tenancy is an architectural pattern to create SaaS solutions. It defines and controls how customers access resources securely and efficiently and how the application is customized per customer easily.

ABP Framework provides a complete multi-tenancy infrastructure. It defines how your application and domain code should be designed, how you access shared resources (such as databases and caches), how you customize the application configuration per customer, and so on. It not only defines but automates wherever possible.

There are two sides of a multi-tenant system:

- **Tenant**: A customer that uses the system and pays for it. A customer has its own users and data, which are isolated from other tenants.

- **Host**: The company that manages the system and the tenants.

You can have separate applications for tenant and host users, or you can build a single application that makes some application features available only for tenant users or only for host users. The ABP startup solution template uses the second approach since it is easier to develop and deploy.

The next section discusses how databases are shared or separated for different tenants.

The database architecture

One of the most fundamental design decisions of a multi-tenant system is how to share or separate the databases of different tenants. There are three common approaches:

- **Single database**: All data of all tenants is stored in a single, shared database. In this case, you should take care when isolating data of different tenants since the database tables are shared.

- **Database (or schema) per tenant**: Every tenant has its own dedicated database. You should dynamically connect to the database for the tenant of the current user.

- **Hybrid**: A mixed approach where some tenants have their own database while others are grouped in one or more databases.

ABP supports the hybrid approach at the framework level by allowing every tenant to have a separate database connection string. However, the startup template and the open source tenant management module come with the single-database model. If you want to use the database per tenant or the hybrid approach, you should customize the tenant management module.

The following figure visualizes the main components of the ABP multi-tenancy infrastructure:

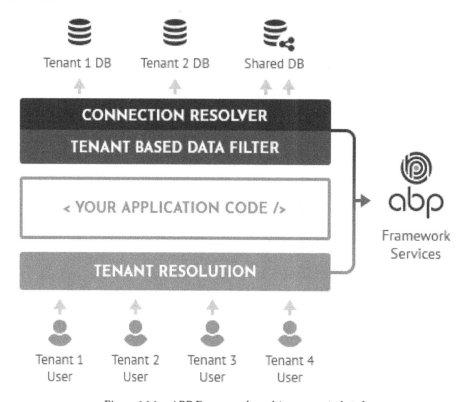

Figure 16.1 – ABP Framework multi-tenancy in brief

ABP Framework's goal is to automate the multi-tenancy-related logic as much as possible and make your application code multi-tenancy-unaware. ABP resolves the current tenant from the HTTP request. It can determine the tenant from the domain (or subdomain) name, cookie, HTTP header, and other parameters. Then, it uses the current tenant information to automatically select the right connection if the tenant has a separate connection string. If the tenant uses a shared database, it automatically filters the data so that a tenant doesn't accidentally access another tenant's data.

We can now start working with the ABP multi-tenancy infrastructure, as we've learned about multi-tenancy and ABP's fundamental multi-tenancy logic overall.

Working with the ABP multi-tenancy infrastructure

In this section, we will explore the basic infrastructure and features of the ABP multi-tenancy system. You will learn how ABP understands the current tenant and isolates the tenant data, how you can get information about the current tenant, and how to switch between tenants. But first, we will start with how you can disable multi-tenancy if you don't need it.

Enabling and disabling multi-tenancy

The ABP startup solution template comes with multi-tenancy enabled by default. The startup solution has a single point that you can use to easily enable or disable multi-tenancy. Find the `MultiTenancyConsts` class inside the `.Domain.Shared` project:

```
public static class MultiTenancyConsts
{
    public const bool IsEnabled = true;
}
```

You can set the `IsEnabled` value to `false` to disable multi-tenancy. This constant is used in a few places in the solution. It is used to set the `AbpMultiTenancyOptions.IsEnabled` option in the `.Domain` project's module class:

```
Configure<AbpMultiTenancyOptions>(options =>
{
    options.IsEnabled = MultiTenancyConsts.IsEnabled;
});
```

ABP uses `AbpMultiTenancyOptions.IsEnabled` to enable or disable multi-tenancy-related features, pages, and components. If you set `MultiTenancyConsts.IsEnabled` to `false` and run the application, you won't see the tenant switch box on the login form and the tenant management page on the main menu anymore. However, the multi-tenancy-related database tables are not removed. The next section explains how to do it.

Removing multi-tenancy tables

Disabling multi-tenancy doesn't remove the multi-tenancy-related database tables from the database. You can leave this as it is (they will already be empty/not used). This way, you can easily enable it for your application later.

If you don't want the multi-tenancy-related tables in your database, find the following line in the `DbContext` class in the `.EntityFramework` project and remove it:

```
builder.ConfigureTenantManagement();
```

Then, remove the implementation of the `ITenantManagementDbContext` interface from your `DbContext` class. You need to remove the `Tenants` and `TenantConnectionStrings` DbSet properties from the class. Finally, remove the `[ReplaceDbContext(typeof(ITenantManagementDbContext))]` attribute from the `DbContext` class declaration. These changes remove the tenant management module's tables from your database schema.

You can add a new database migration to remove the tables from the database. Run the following command in the root directory of the `.EntityFramework` project:

```
dotnet ef migrations add Removed_TenantManagement
```

Then, run the following command to apply changes to the database:

```
dotnet ef database update
```

In this way, your database won't include multi-tenancy-related tables. You can also remove the `Volo.Abp.TenantManagement.*` NuGet packages from the projects in the solution and the code parts using these packages. However, all these are optional. I suggest you keep them if you think you may enable multi-tenancy for your application later because they have no functionality as long as the `AbpMultiTenancyOptions.IsEnabled` option is set to `false`.

As you've seen, enabling/disabling multi-tenancy with ABP Framework is just a single line of change. If you decide to develop your application as multi-tenancy-enabled, you can continue with the next section to understand how ABP determines the current tenant from HTTP requests.

Determining the current tenant

If you look at *Figure 16.1* again, you will see that all the requests coming from the users are passing through the tenant resolution component before executing the application code. This way, the current tenant becomes known inside your application.

Intercepting the incoming requests is done with ABP's multi-tenancy middleware component. All the hosting projects in the startup solution template contain the following lines in the `OnApplicationInitialization` method of the ABP module class:

```
if (MultiTenancyConsts.IsEnabled)
{
    app.UseMultiTenancy();
}
```

This middleware is added after the authentication middleware (because the user's authentication ticket is used on tenant resolution) and before the authorization middleware (because ABP authorizes users based on their tenants).

The multi-tenancy middleware resolves the current tenant from the HTTP request and sets the `ICurrentTenant` properties that are used to obtain the current tenant information. The `ICurrentTenant` interface will be explained in the next section, but we should first understand how ABP determines the current tenant from the HTTP request.

The current tenant information is obtained from the current HTTP request using request parameters in the following order:

1. If the user (or client) has authenticated, then the current tenant's ID and name are extracted from the claims in the authentication ticket (either in the cookie or in the header, based on the authentication method).

2. If `AbpTenantResolveOptions` is configured, the tenant's name is determined from the domain (or subdomain) name.

3. The `__tenant` query string parameter is used to get the tenant's name or ID if the current HTTP request contains that parameter.

4. The `__tenant` route parameter is used to get the tenant's name or ID if the current HTTP request contains that parameter.

5. The `__tenant` HTTP header is used to get the tenant's name or ID if the current HTTP request contains that parameter.

6. The `__tenant` cookie's value is used to get the tenant's name or ID if the current HTTP request contains that parameter.

If ABP determines the tenant in any of the preceding steps, it doesn't continue to other steps as you might expect. If none of the information is found in the HTTP request, then it is assumed that the current user is a host user. All the options are already preconfigured and working when you create a new solution, so you typically do not make many configurations for your solution. You should only care about the domain name resolution, which is suggested in the production environment.

The following example shows how to configure the domain name resolver in the `ConfigureServices` method of your module class:

```
Configure<AbpTenantResolveOptions>(options =>
{
    options.AddDomainTenantResolver("{0}.yourdomain.com");
});
```

The `AddDomainTenantResolver` method accepts a domain format where the `{0}` part matches the tenant name. This means if your tenant's name (the `Name` property of the `Tenant` class) is `acme`, then the `acme` users should use the `acme.yourdomain.com` URL to enter the application.

Once ABP resolves the tenant, we can work with the current tenant, as explained in the next section.

Working with the current tenant

ABP determines the tenant before executing our application code, as we've learned in the previous section. We can get the current tenant's information using the `ICurrentTenant` service. The following example demonstrates how to use the `ICurrentTenant` service in an arbitrary class:

```
public class MyService : ITransientDependency
{
    private readonly ICurrentTenant _currentTenant;

    public MyService(ICurrentTenant currentTenant)
    {
        _currentTenant = currentTenant;
    }

    public async Task DoItAsync()
    {
```

```
            Guid? tenantId = _currentTenant.Id;
            string tenantName = _currentTenant.Name;
        }
    }
```

We've injected the ICurrentTenant service and accessed the Id and Name properties in the example method. The Id and Name properties return null if the current user is a host user (which means that the tenant is not available). Some ABP base classes already pre-inject the ICurrentTenant service, so you can directly use the CurrentTenant property, as shown in the following example:

```
public class MyAppService : ApplicationService
{
    public async Task DoItAsync()
    {
        Guid? tenantId = CurrentTenant.Id;
    }
}
```

Since the ApplicationService base class already has the CurrentTenant property (of the ICurrentTenant type), we can use it directly without manual injection.

ICurrentTenant has no more important properties. If you need to get more information/data for the current tenant, you have the tenant's Id property to query from the database.

Most of the time, your application code will work with the current tenant. But sometimes, you may need to change the current tenant, as explained in the next section.

Switching between tenants

The ICurrentTenant service is also used by ABP Framework to isolate the current tenant's data automatically so that you don't accidentally access other tenant data. However, in some cases, you may need to work with another tenant's data in the same HTTP request and temporarily switch the tenant. The ICurrentTenant service is not only used to get information about the current tenant but also to switch to the desired tenant. See the following example:

```
public class MyAppService : ApplicationService
{
    public async Task DoItAsync(Guid tenantId)
```

```
    {
        // Before the using block
        using (CurrentTenant.Change(tenantId))
        {
            // Inside the using block
            // CurrentTenant.Id equals to tenantId
        }
        // After the using block
    }
}
```

If you use the CurrentTenant.Id property before the using block, you get the tenant's ID that was resolved, as explained in the *Determining the current tenant* section. The CurrentTenant.Change method changes the current tenant to a given value, so you get the desired tenant's ID when using the CurrentTenant.Id property inside the using block. For example, if you perform a database query from a shared database inside the using block, ABP will retrieve the desired tenant's data instead of the one resolved by the multi-tenancy middleware. Once the using block completes, CurrentTenant.Id is automatically restored to the previous value. You can safely use the CurrentTenant.Change method in a nested way when you rarely need it. If you want to switch to the host context, you can pass a null value to the Change method. Always use the Change method with a using block, as in this example, to not affect the surrounding context of your method.

In addition to switching to the desired tenant, it is also possible to completely disable the tenant isolation.

Disabling the data isolation

Data isolation is critical in a multi-tenant application. It guarantees to query only the current tenant's data. However, in some cases, your application may require querying from the entire database, including all tenants' data.

We explored the ABP's data filtering system in the *Using the data filtering system* section of *Chapter 8, Using the Features and Services of ABP*. ABP uses the same data-filtering system to filter data of the current tenant. So, we can use the same data-filtering API to disable the multi-tenancy filter temporarily:

```
public class ProductAppService : ApplicationService
{
    private readonly IRepository<Product, Guid>
```

```
        _productRepository;

    public ProductAppService(
        IRepository<Product, Guid> productRepository)
    {
        _productRepository = productRepository;
    }

    public async Task<long> GetTotalProductCountAsync()
    {
        using (DataFilter.Disable<IMultiTenant>())
        {
            return await
                _productRepository.GetCountAsync();
        }
    }
}
```

In this example, we are getting the total number of products owned by all the tenants in the database. The multi-tenancy data filter is disabled in the using block, so the repository works with all the records in the database.

While disabling the multi-tenancy filter is pretty easy, there is an important limitation – it only works with the single database approach. It cannot query a tenant's data if the tenant has a dedicated database. Currently, there is no direct way to perform a query on multiple databases and aggregate the query results as a single result set.

Besides the technical limitation, there is also a design problem with querying all tenants' data. Ideally, multi-tenant software should be designed so that all tenants have their on-premises deployment with separate database and application servers. We will return to this discussion later in the *When to use multi-tenancy* section of this chapter.

We've learned the ways to access and change the current tenant. In the next section, we will see how to design our entities to be multi-tenancy-compatible.

Designing the domain as multi-tenant

ABP aims to make your application code multi-tenancy-unaware and automate things wherever possible. Designing an entity class as multi-tenant is very simple. Just implement the `IMultiTenant` interface for your entity, as shown in the following example:

```
public class Product : AggregateRoot<Guid>, IMultiTenant
{
    public Guid? TenantId { get; set; }
    public string Name { get; set; }
}
```

The `Product` aggregate root entity in this example implements the `IMultiTenant` interface and defines a `TenantId` property. The tenant identifier type is always `Guid` in ABP Framework. The `TenantId` property is nullable, making the `Product` entity available both for the tenant and the host side. If the `TenantId` property is `null`, that entity belongs to the host side. It also allows us to easily convert our application to a single-tenant, on-premises application where the `TenantId` property is always `null`.

ABP automatically sets the `TenantId` value using the `ICurrentTenant.Id` property when you create a new entity object (a `Product` object for this example). ABP is also responsible for saving it to the right database and querying from the right database, or filtering the tenant data if you are using a single database.

We've learned the fundamental points of building a multi-tenant solution with ABP Framework. The next section introduces the ABP feature system that can be used to restrict application functionalities for tenants.

Using the feature system

Most SaaS solutions provide different packages to the customers. Every package has a different set of application features and is subscribed at a different price. ABP provides a feature system used to define such application features, and then disable or enable these features for individual tenants. Let's start by defining a feature.

Defining the features

It is required to define a feature before using it. Create a new class deriving from the
`FeatureDefinitionProvider` class (typically in the `.Application.Contracts`
project in the startup solution) and override the `Define` method, as shown in the
following example:

```
public class MyAppFeatureDefinitionProvider :
    FeatureDefinitionProvider
{
    public override void Define(
        IFeatureDefinitionContext context)
    {
        var myGroup = context.AddGroup("MyApp");
        myGroup.AddFeature(
            "MyApp.StockManagement",
            defaultValue: "false",
            displayName: L("StockManagement"),
            isVisibleToClients: true);
        myGroup.AddFeature(
            "MyApp.MaxProductCount",
            defaultValue: "100",
            displayName: L("MaxProductCount"));
    }

    private ILocalizableString L(string name)
    {
        return
            LocalizableString.Create<MtDemoResource>(name);
    }
}
```

Features are grouped to create more modular systems (where every module defines its
own group). In this example, I created a feature group for the final application. Then, I
defined two features under that group. This example defines the two features:

- The first one is used to enable or disable the stock management feature for tenants.

- The second one is used to limit the product entity count.

Feature values are actually just strings, such as `false` and `100` in this example. However, Boolean values (`true` and `false`) can be used for conditional checks by convention.

ABP automatically discovers classes derived from the `FeatureDefinitionProvider` class, so you don't need to register it somewhere. After defining a feature, we can check its value for the current tenant (we will see how to assign features to tenants later, in the *Managing tenant features* section).

Checking for the features

I will show you how to enable, disable, or set the value of a feature for a tenant in the *Managing tenant features* section. But first, I want to show you how we check a feature's value for a tenant.

There are several ways to check a feature's value for the current tenant. The easiest way is to use the `RequiresFeature` attribute that can be used on methods or classes.

Using the RequiresFeature attribute

The following example uses the `RequiresFeature` attribute to restrict a class's usage for the current tenant:

```
[RequiresFeature("MyApp.StockManagement")]
public class StockAppService : ApplicationService,
    IStockAppService
{
}
```

This way, the `MyApp.StockManagement` feature's value is automatically checked in every method call for the `StockAppService` service, and an exception is thrown for unauthorized access.

The `RequiresFeature` attribute can also be used on a method. See the following example:

```
public class ProductAppService : ApplicationService
{
    . . .
    [RequiresFeature("MyApp.StockManagement")]
    public async Task<long> GetStockCountAsync()
    {
        return await _productRepository.GetCountAsync();
```

```
    }
}
```

In this case, only the `GetStockCountAsync` method is restricted, and the other methods of `ProductAppService` without the `RequiresFeature` attribute are not affected.

The `RequiresFeature` attribute is easy to use but limited to Boolean features (with `true` and `false` values). For detailed usage, we should use the `IFeatureChecker` service.

Using the IFeatureChecker service

The `IFeatureChecker` service allows us to get and check the feature values programmatically. You can inject it just like any other service. The following example checks whether the `MyApp.StockManagement` feature is enabled for the current tenant:

```
public async Task<long> GetStockCountAsync()
{
    if (await FeatureChecker
            .IsEnabledAsync("MyApp.StockManagement"))
    {
        return await _productRepository.GetCountAsync();
    }
    // TODO: Your fallback logic or error message
}
```

The `IsEnabled` method returns `true` only if the feature's value is `true` (as `string`). If you have fallback logic (when the tenant doesn't have that feature enabled), then using `IsEnabledAsync` is a good approach. However, if you only want to check whether a feature is enabled and should throw an exception otherwise, use the `CheckEnabledAsync` method, as shown in the following example:

```
public async Task<long> GetStockCountAsync()
{
    await FeatureChecker.CheckEnabledAsync("MyApp.
StockManagement");
    return await _productRepository.GetCountAsync();
}
```

The `CheckEnabledAsync` method throws `AbpAuthorizationException` if the given feature is not enabled for the current tenant. However, if you need to simply check whether a feature is enabled or disabled at the beginning of the method, using the `RequiresFeature` attribute would be simpler.

Using the `IFeatureChecker` service is especially useful when you want to get the value of non-Boolean features. For example, the `MyApp.MaxProductCount` feature introduced in the *Defining the features* section is a numeric feature. We can't simply check whether it is enabled or disabled. We need to know its value for the current user.

The following example checks the maximum allowed product count for the current tenant before creating a new product:

```
public async Task CreateAsync(string name)
{
    var currentProductCount = await
        _productRepository.GetCountAsync();
    var maxProductCount = await
            FeatureChecker.GetAsync<int>(
                "MyApp.MaxProductCount");
    if (currentProductCount >= maxProductCount)
    {
        // TODO: Throw a business exception
    }
    // TODO: Continue to create the product
}
```

The `FeatureChecker.GetAsync<T>` method returns the value of the given feature by converting to the given generic type argument. Here, `MyApp.MaxProductCount` is a numeric feature, so I am converting to `int` and then comparing it with the current product count of the current tenant. `IFeatureChecker` also defines the `GetOrNullAsync` method that returns the string value of the feature or returns `null` if the feature has no value defined for the current tenant.

Checking the Feature of Another Tenant

The `IFeatureChecker` service works for the current tenant. If you want to check a feature's value for another tenant when you have the other tenant's ID, first switch to the target tenant, as explained in the *Switching between tenants* section, and then use the `IFeatureChecker` service as normal.

The `RequiresFeature` attribute and the `IFeatureChecker` service are used on the server side, but we also need to get and check feature values in our client applications.

Checking features on the client side

When you define a feature, you will need to know its value on the client side. For example, if the `MyApp.StockManagement` feature is not enabled for the current tenant, you generally want to hide the related UI elements from the application pages and disable the client-to-server HTTP API calls for this feature.

ABP provides multiple UI options, and each one provides a different API to check the features on the client side. For example, the ABP MVC/Razor Pages UI provides the global `abp.features` JavaScript API to check the features, as shown in the following code block:

```
if (abp.features.isEnabled('MyApp.StockManagement'))
{
    // TODO: ...
}
```

Please see the *Checking the tenant features* section of *Chapter 12*, *Working with MVC/ Razor Pages*, for more details about the `abp.features` JavaScript API.

The ABP Blazor UI, on the other hand, uses the same `IFeatureChecker` service on the client side. For other UI types, please refer to ABP's documentation: `https://docs. abp.io/en/abp/latest/Features`.

You've now learned how to get and check a feature's value for the current tenant. The next section explains how you can set the value of a feature for a tenant.

Managing tenant features

At a framework level, ABP doesn't care where the feature values are stored and how they are changed. It just defines an interface, `IFeatureStore`, that can be implemented to obtain the current value of a feature. However, it would not be good to leave its implementation to every developer because, most of the time, the implementation will be similar, and we don't want to waste our time re-implementing it again and again.

ABP Framework provides the **Feature Management** module, which implements the `IFeatureStore` interface and provides a UI and API to modify the feature values for the tenants. The Feature Management module is already installed when you create a new solution using ABP's startup solution template. The following sections explain the Feature Management UI modal and the API to manage the feature values.

Using the Feature Management UI modal

The Feature Management module can automatically create the modal dialog to set the feature values. However, we need to define the value type for each feature. Return to `MyFeatureDefinitionProvider` and update the feature definitions as follows:

```
myGroup.AddFeature(
    "MyApp.StockManagement",
    defaultValue: "false",
    displayName: L("StockManagement"),
    isVisibleToClients: true,
    valueType: new ToggleStringValueType());
myGroup.AddFeature(
    "MyApp.MaxProductCount",
    defaultValue: "100",
    displayName: L("MaxProductCount"),
    valueType: new FreeTextStringValueType(
                new NumericValueValidator()));
```

I added `valueType` parameters to the `AddFeature` methods. The first one is `ToggleStringValueType`, which indicates that the feature has an on/off style (Boolean) value. The second one is `FreeTextStringValueType`, which indicates that the feature has a value that should be changed with a textbox. `NumericValueValidator` specifies the validation rule for the value.

Once we have defined the value types properly, the Feature Management module can automatically render the necessary UI to set the feature values. To open the Feature Management dialog, log in to the application as an authorized host user, navigate to the tenant management page from the main menu, click the **Actions** button, and select the **Features** action, as shown in the following figure:

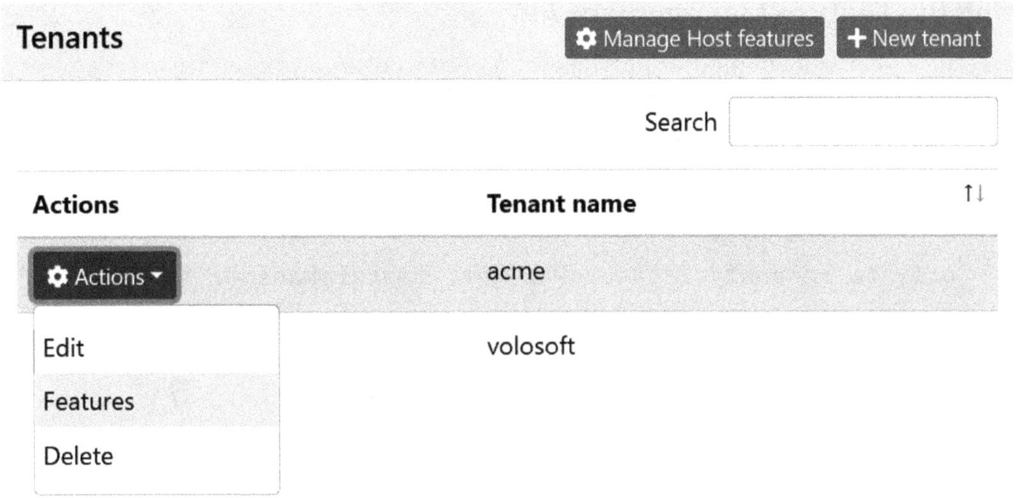

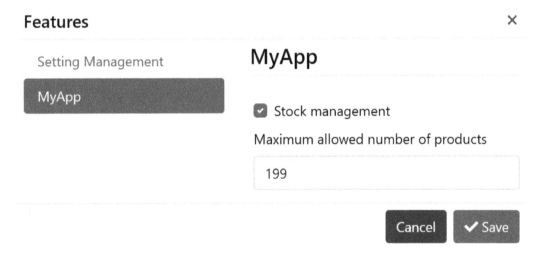

Figure 16.2 – The Features action on the tenant management page

This action will open a modal dialog, as shown in the following figure:

Figure 16.3 – The Feature Management dialog

We can see the group name on the left side (you can also localize the group's display name). When we click the **MyApp** group, we can see the form elements to set the values of the features. The UI has been dynamically created by the Feature Management module.

The MyApp.StockManagement feature becomes a checkbox on the UI, while a numeric textbox is shown for the MyApp.MaxProductCount feature. This way, you can easily set the feature values for any tenant. In addition to the UI, it is also possible to programmatically set the feature values using the Feature Management API.

Using the Feature Management API

The Feature Management module provides the `IFeatureManager` service to set a feature's value for a tenant programmatically. The following example enables the `MyApp.StockManagement` feature for the given tenant:

```
public class MyCustomerService : DomainService
{
    private readonly IFeatureManager _featureManager;

    public MyCustomerService(IfeatureManager
                            featureManager)
    {
        _featureManager = featureManager;
    }

    public async Task EnableStockManagementAsync(Guid
                                                 tenantId)
    {
        await _featureManager.SetForTenantAsync(
            tenantId,
            "MyApp.StockManagement",
            "true"
        );
    }
}
```

We are injecting the `IFeatureManager` service into our class's constructor, just like any other service. Then, we use the `SetForTenantAsync` method to set the value to `true` for the given tenant.

When to use multi-tenancy

Multi-tenancy is a great pattern to create SaaS solutions, and ABP Framework provides a complete infrastructure to create multi-tenant applications. However, not all applications should be SaaS, and not all SaaS applications should be multi-tenant. ABP's multi-tenancy system has some assumptions, and we've made some design decisions while building it. In this section, I want to talk about these assumptions and decisions to help you to decide whether ABP's multi-tenancy system fits into your solution.

ABP multi-tenant applications should be developed by assuming that each tenant will have a separated and isolated production environment. If you make this assumption, then you will have some restrictions. Here are a few example restrictions:

- You should not perform database queries from multiple tenants at once. If you do this, you assume that you will have a shared tenant database because it is technically not straightforward to query from multiple databases from different (and probably isolated) environments.

- A tenant's user cannot log in to the system with another tenant. This means you can't assign multiple tenants to a single user since users are completely isolated. ABP allows using the same email address or user name in different tenants, but they will actually be different users with different passwords and identifiers in the database, without any relation.

- You cannot share roles (and their permissions) between different tenants.

These are natural restrictions if you assume that two tenants have different production environments and can't access each other's environment. ABP assumes that the same application can be deployed on-premises for a customer without any code change (except the `AbpMultiTenancyOptions.IsEnabled` option).

These assumptions don't mean that the tenants cannot share data at all. If an entity doesn't implement the `IMultiTenant` interface, it is naturally shared among all tenants and always stored in the central (host) database. In addition, you can switch between tenants to temporarily access a tenant's data from another tenant user. However, you should think about how this logic will work in an on-premises environment, or you can drop the on-premises deployment support from your solution.

Most of the confusion comes from thinking about the multi-tenancy only from a technical perspective but not thinking about its purpose of design. For example, think of an electronic marketplace where vendors manage and sell their products. Individual customers list and search products, add to a cart, and make a payment. This application may seem like a multi-tenant system if you assume that the vendors have their own products and vendor back-office users manage these products. If you use ABP's multi-tenancy system, all the isolation will be automatically done, right?

While it has some requirements similar to a multi-tenant system from a technical perspective, the marketplace is likely to have parts that are integrated as a unified platform. In a multi-tenant system, a customer (tenant) behaves as if it owns the entire system. In a marketplace, a vendor is not a tenant. It does not use an application in isolation, as in an on-premises system. So, if you start with multi-tenancy, you will later have to deal with the problems of data sharing and integrations because shared/integrated parts are much more than the isolated parts in such a system.

Summary

In this chapter, we've explored the fundamental infrastructure provided by ABP Framework. We've learned how ABP determines the current tenant and isolates the data between the tenants. We also learned how to switch to another tenant or completely disable the data isolation when needed.

Another great ABP feature is the feature system. We've defined features by creating a feature provider class and learned different ways of checking the value of a feature for the current user.

You are now able to work on a multi-tenant application's development where the tenants can have different rights on the application features.

The next chapter introduces different levels of automated testing and explains how you can create unit and integration tests for your ABP-based solutions.

17
Building Automated Tests

Building automated tests is an essential `GetRegistrationOrNull` practice to create maintainable software solutions and is a fast and repeatable way of validating the software. ABP Framework and the ABP startup solution template are designed with testability in mind. We've already seen an example of writing a simple integration test with ABP Framework in *Chapter 3, Step-By-Step Application Development*.

In this chapter, you will understand the ABP test infrastructure and build unit and integration tests for your ABP-based solutions. You will learn about data seeding for tests, mocking the database, and testing different kinds of objects. You will also learn the basics of automated tests such as assertions, mocking and replacing services, and dealing with exceptions.

Here is a list of the main topics covered in this chapter:

- Understanding the ABP test infrastructure
- Building unit tests
- Building integration tests

Technical requirements

If you want to follow the examples in this chapter, you need to have an IDE/editor that supports ASP.NET Core development.

The examples in this chapter are mostly based on the EventHub solution I introduced in *Chapter 4, Understanding the Reference Solution.* Please refer to that chapter to learn how to download the source code of the EventHub solution.

Understanding the ABP test infrastructure

ABP's startup solution template includes preconfigured test projects to build unit and integration tests for your solution. While you can write your tests without understanding the complete structure, I think it is worth exploring this so that you can understand how it works and customize it when you need it. We'll start by exploring the test projects.

Exploring the test projects

The following screenshot shows the test projects that get created when you create a new ABP solution:

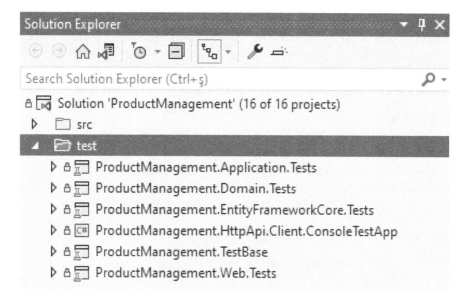

Figure 17.1 – Test projects in the ABP startup solution

The preceding screenshot shows the `test` projects for a solution named `ProductManagement`, with the MVC/Razor Pages UI and the **Entity Framework Core** (**EF Core**) database provider. The `test` project list may be slightly different if you use a different UI or database provider, but the fundamental logic is the same. The following list explains the projects in general:

- `ProductManagement.HttpApi.Client.ConsoleTestApp`: A very simple console application to manually test the HTTP API endpoints of your application. So, this is not a part of our automated test infrastructure, and you can ignore it for this chapter.

- `ProductManagement.TestBase`: A project that is shared by the other test projects. It has references to the base test libraries and includes data seeding and some other base configuration code. It doesn't contain any test class normally.

- `ProductManagement.EntityFrameworkCore.Tests`: You can build tests for EF Core integration code in this project, such as your custom repositories. This project also configures a SQLite in-memory database for your tests.

- `ProductManagement.Domain.Tests`: Use this project to build tests for your domain layer.

- `ProductManagement.Application.Tests`: Use this project to build tests for your application layer.

- `ProductManagement.Web.Tests`: Use this project to build tests for your MVC/Razor Pages UI.

The solution uses some libraries as the test infrastructure, as explained in the next section.

Exploring the test libraries

The `ProductManagement.TestBase` project has reference to the following NuGet packages:

- `xunit`: xUnit is one of the most popular test frameworks for .NET.

- `Shouldly`: A library to write the assertion code in an easy and readable format.

- `NSubstitute`: A library to mock objects in unit tests.

- `Volo.Abp.TestBase`: ABP's package to easily create ABP-integrated test classes.

We will see how to use the basics of these libraries in the *Building unit tests* and *Building integration tests* sections. Before starting to write our tests, let's see how we can run the tests.

Running the tests

In this section, I will show two ways of running the tests. The first way is to use an IDE that supports running test execution. I will use Visual Studio as an example. You can open the **Test Explorer** window from the **Test | Test Explorer** item on the main menu:

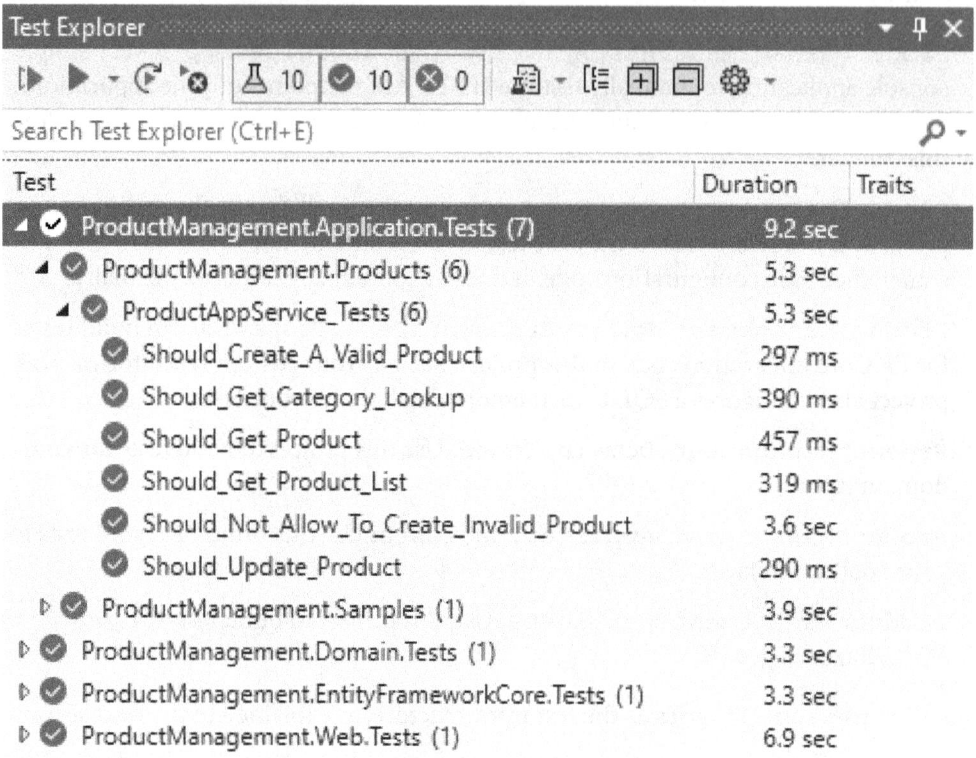

Figure 17.2 – Test Explorer in Visual Studio

Test Explorer automatically discovers all the tests in your solution and allows you to run some or all of them. In *Figure 17.2*, I ran all the tests, and all of them succeeded. This screenshot has been taken from the `ProductManagement` application built in *Chapter 3, Step-By-Step Application Development*, and the source code can be found at `https://github.com/PacktPublishing/Mastering-ABP-Framework`.

Visual Studio runs the tests one by one by default, and thus it takes a long time to run all the tests. You can click the down arrow icon near the cog icon in **Text Explorer** and select the **Run Tests In Parallel** option (see *Figure 17.3*) to run the tests in parallel so that it takes significantly less time to run them all:

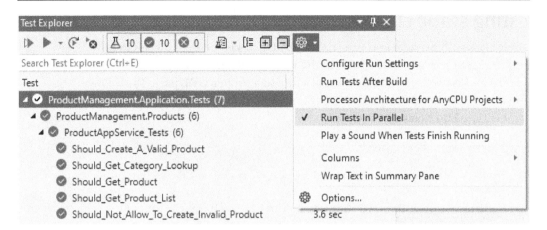

Figure 17.3 – Running tests in parallel in Visual Studio

ABP Framework and the startup solution template have been designed to support running tests in parallel so that tests don't affect each other.

An alternative way to run tests is to use the `dotnet test` command in the root directory of your solution. It automatically discovers and runs all the tests and reports the test result in the command-line terminal. This command exits with a 0 (success) return code if all the tests succeed; otherwise, if any test fails, it exits with a 1 return code. This command is especially useful if you build a **continuous integration** (**CI**) pipeline where you automatically run the tests.

You've learned the test structure of the ABP startup solution and have run the automated tests. Now, we can start to build our tests.

Building unit tests

In this section, we will see different types of unit tests. We will begin by testing a static class, then we will write tests for a class with no dependencies. We will continue with a class with dependent services and learn how to mock these dependencies to unit test that class. We will learn the basics of writing automated test code with examples.

Let's begin with the simplest case—testing static classes.

Testing static classes

A static class with no state and external dependencies is the easiest class to test. `EventUrlHelper` is a static class (in the `EventHub.Domain` project of the EventHub solution) and is used to convert an event's title to a proper URL part. The following test class (in the `EventHub.Domain.Tests` project of the EventHub solution) tests the `EventUrlHelper` class:

```
public class EventUrlHelper_Tests
{
    [Fact]
    public void Should_Convert_Title_To_Proper_Urls()
    {
        var url = EventUrlHelper.ConvertTitleToUrlPart(
                    "Introducing ABP Framework!");
        Assert.Equal("introducing-abp-framework", url);
    }
}
```

The first rule is that the test class should be `public`. Otherwise, you can't see it in **Test Explorer**. The `[Fact]` attribute is defined by the xUnit library. Any public method with the `[Fact]` attribute is considered a test case and is automatically discovered by **Test Explorer**. You can give your method whatever name you want, but the suggested naming pattern is reflected in this example. You can even name it more specifically, such as `Should_Convert_Url_To_Kebab_Case`, and only test the functionality related to kebab-case.

The test code in this example is very simple. We call the static `EventUrlHelper.ConvertTitleToUrlPart` method with a sample title value, then compare the result with the value we expect it to be. The `Assert` class is defined by xUnit, with many methods to define our expectations. The test case succeeds only if the given values are equal. Otherwise, we see a red icon for the test case in **Test Explorer** with an error message indicating what's wrong with the test.

You can right-click on a specific test in **Test Explorer** to run it and see the result, as depicted in the following screenshot:

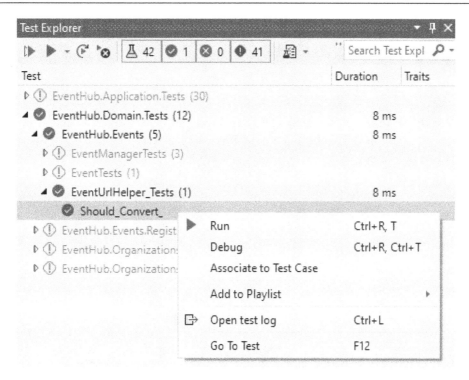

Figure 17.4 – Running a specific test in Test Explorer in Visual Studio

Another common xUnit attribute is [Theory], which provides parameters to a test method and tests it for each parameter set. Assuming that we want to run the test with different event URLs, we can rewrite the test method, as shown in the following code block:

```
public class EventUrlHelper_Tests
{
    [Theory]
    [InlineData("Introducing ABP Framework!",
                "introducing-abp-framework")]
    [InlineData("Blazor: UI Messages",
                "blazor-ui-messages")]
    [InlineData("What's new in .NET 6",
                "whats-new-in-net-6")]
    public void Should_Convert_Title_To_Proper_Urls(
        string title, string url)
    {
        var result =
```

```
                    EventUrlHelper.ConvertTitleToUrlPart(title);
            result.ShouldBe(url);
    }
}
```

xUnit runs this test method for each `[InlineData]` set separately and passes the `title` and `url` parameters as the given data. If you look at **Test Explorer** again, you will see these three test cases there:

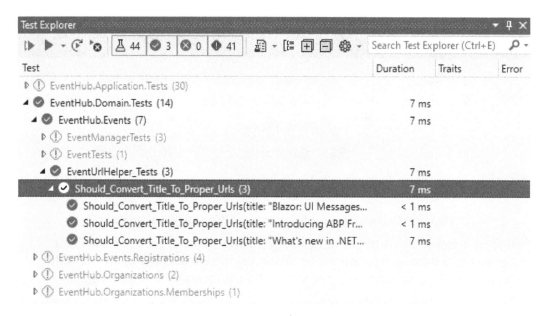

Figure 17.5 – Using the [Theory] attribute for unit tests

I also used the `Shouldly` library for the assertion in this example. The `result.ShouldBe(url)` expression is simpler to write and read than the `Assert.Equal(url, result)` expression. The `Shouldly` library works with extension methods such as that, and I will use it in future examples.

Testing static classes (with no state and external dependencies) was easy. We've also learned some xUnit and `Shouldly` features. The next section continues with testing simple classes without a service dependency.

Testing classes with no dependencies

Some classes, such as entities, may have no dependency on other services. Testing these classes is relatively easy since we don't need to prepare the dependencies to make the class work properly.

The following test method tests the `Event` class's constructor:

```
public class Event_Tests
{
    [Fact]
    public void Should_Create_A_Valid_Event()
    {
        new Event(
            Guid.NewGuid(),
            Guid.NewGuid(),
            "1a8j3v0d",
            "Introduction to the ABP Framework",
            DateTime.Now,
            DateTime.Now.AddHours(2),
            "In this event, we will introduce the ABP
            Framework..."
        );
    }
}
```

In this example, I've passed a valid list of parameters so that it doesn't throw an exception, and the test succeeds. The following example tests for an exception case:

```
[Fact]
public void
    Should_Not_Allow_End_Time_Earlier_Than_Start_Time()
{
    var exception = Assert.Throws<BusinessException>(() =>
    {
        new Event(
            Guid.NewGuid(),
            Guid.NewGuid(),
            "1a8j3v0d",
            "Introduction to the ABP Framework",
            DateTime.Now, // Start time
            DateTime.Now.AddDays(-2), // End time
            "In this event, we will introduce the ABP
            Framework..."
```

```
        );
    });


    exception.Code.ShouldBe(EventHubErrorCodes
        .EventEndTimeCantBeEarlierThanStartTime);
}
```

I intentionally passed the end time as 2 days earlier than the start time. I am expecting the constructor to throw a BusinessException exception by using the Assert. Throws<T> method. If the code block inside the Throws method throws an exception of type BusinessException, then the test passes; otherwise, the test will fail. I am also checking the error code with the ShouldBe extension method.

Let's write a method that tests another method of the Event class. The following example creates a valid Event object, then changes its start and end times, and finally checks whether the times were changed:

```
[Fact]
public void Should_Update_Event_Time()
{
    // ARRANGE
    var evnt = new Event(
        Guid.NewGuid(),
        Guid.NewGuid(),
        "1a8j3v0d",
        "Introduction to the ABP Framework",
        DateTime.Now,
        DateTime.Now.AddHours(2),
        "In this event, we will introduce the ABP
        Framework..."
    );
    var newStartTime = DateTime.Now.AddHours(1);
    var newEndTime = DateTime.Now.AddHours(2);


    //ACT
    evnt.SetTime(newStartTime, newEndTime);


    //ASSERT
```

```
        evnt.StartTime.ShouldBe(newStartTime);
        evnt.EndTime.ShouldBe(newEndTime);
        evnt.GetLocalEvents()
            .ShouldContain(x => x.EventData
                            is EventTimeChangingEventData);
}
```

This example fully implements the common **Arrange-Act-Assert** (**AAA**) test pattern, detailed as follows:

- The *Arrange* part prepares the objects we need to work on.

- The *Act* part executes the actual code we want to test.

- The *Assert* part checks whether the expectations are met.

I suggest separating your test method's body with these comment lines to make what you are testing and asserting explicit. In this example, we used the SetTime method of the Event class to change the event times. The SetTime method also publishes a local event, so I checked it too in the *Assert* part.

As you see in the examples, if the class we want to test has no external dependencies, we can simply create an instance and execute the methods on it. In the next section, we will see how to deal with external dependencies.

Testing classes with dependencies

Most services have dependencies on other services. We use the **dependency injection** (**DI**) system to take these dependencies into the service's constructor. The purpose of unit testing is to test a class as isolated from other classes because unit tests should generally have only one reason to fail. We should somehow exclude the dependencies while testing the target class. In this way, our test is affected by the changes in the target class but not affected by changes in other classes.

Mocking is a technique used in unit testing to replace a target class's dependencies with fake implementations so that the test isn't affected by the target class's dependencies.

I will test the IsPastEvent method of the EventRegistrationManager class as an example. The IsPastEvent method gets an event.

EventRegistrationManager is a domain service and takes three external services in its constructor, as shown in the following simplified code block:

```
public class EventRegistrationManager : IDomainService
{
    ...
    public EventRegistrationManager(
        IEventRegistrationRepository
            eventRegistrationRepository,
        IGuidGenerator guidGenerator,
        IClock clock)
    {
        _eventRegistrationRepository =
            eventRegistrationRepository;
        _guidGenerator = guidGenerator;
        _clock = clock;
    }

    public bool IsPastEvent(Event @event)
    {
        return _clock.Now > @event.EndTime;
    }
}
```

We should pass instances of these three external services to be able to create an EventRegistrationManager object. The following code block shows how I've written a test method for the IsPastEvent method of that class:

```
public class EventRegistrationManager_UnitTests
{
    [Fact]
    public void IsPastEvent()
    {
        var clock = Substitute.For<IClock>();
        clock.Now.Returns(DateTime.Now);

        var registrationManager = new
            EventRegistrationManager(null, null, clock
```

```
        );

    var evnt = new Event(
        Guid.NewGuid(),
        Guid.NewGuid(),
        "1a8j3v0d",
        "Introduction to the ABP Framework",
        DateTime.Now.AddDays(-10), // Start time
        DateTime.Now.AddDays(-9), // End time
        "In this event, we will introduce the ABP
            Framework..."
    );

    registrationManager.IsPastEvent(evnt)
        .ShouldBeTrue();
    }
}
```

The test code starts by creating a fake `IClock` object using the `Substitute.For<T>` utility method of the `NSubstitute` library. The `clock.Now.Returns(DateTime.Now)` statement configures the fake object so that it returns `DateTime.Now` whenever the `clock.Now` property is called. We do that since the `IsPastEvent` method will call the `clock.Now` property. That means we should know the internal implementation details of the unit-tested method to test it properly.

Since I know that the `IsPastEvent` method won't use the `IEventRegistrationRepository` and `IGuidGenerator` services, I can pass them as `null` in the constructor of the `EventRegistrationManager` class.

Finally, I've called the `IsPastEvent` method of the `EventRegistrationManager` class with an example event and checked the result.

Let's see a more complex example. This time, we are testing the `RegisterAsync` method of the `EventRegistrationManager` class. The code is illustrated in the following snippet:

```
[Fact]
public async Task
    Valid_Registrations_Should_Be_Inserted_To_Db()
{
```

```
    var evnt = new Event(/* some valid arguments */);
    var user = new IdentityUser(/* some valid arguments
                          */);

    var repository =
        Substitute.For<IEventRegistrationRepository>();
    repository
        .ExistsAsync(evnt.Id, user.Id)
        .Returns(Task.FromResult(false));
    var clock = Substitute.For<IClock>();
    clock.Now.Returns(DateTime.Now);
    var guidGenerator = SimpleGuidGenerator.Instance;

    var registrationManager = new EventRegistrationManager(
        repository, guidGenerator, clock
    );

    await registrationManager.RegisterAsync(evnt, user);

    await repository
        .Received()
        .InsertAsync(
            Arg.Is<EventRegistration>(
            er => er.EventId == evnt.Id && er.UserId ==
                user.Id)
    );
}
```

First, I've created an Event object and an IdentityUser object because
the RegisterAsync method gets these parameters. Then, I've mocked the
EventRegistrationManager dependencies. Since the RegisterAsync
method uses all the dependencies, I had to mock them all. See how I configured the
fake repository to return false when the ExistsAsync method is called. The
RegisterAsync method uses the ExistsAsync method to check whether there is
already a registration with the same event and user.

After executing the `RegisterAsync` method, I should somehow check whether the registration is complete. I can use the `Received` method of `NSubstitute` to check whether the repository's `InsertAsync` method is called with an `EventRegistration` object with the specified event and **user identifiers** (**UIDs**).

In this section, I've covered the basics of unit testing. Unit tests have two main advantages compared to integration tests, as outlined here:

- They run fast because only the tested class really works. All others are mocked and typically have no execution cost.

- They make it easier to investigate problems. If a class doesn't work properly, only the tests working on that class fail, so you can easily find the problem's source.

However, writing and maintaining unit tests is hard when your classes have dependencies. Unit tests also can't tell much about whether your class will properly work at runtime integrated with other services. That brings us to integration tests.

Building integration tests

In this section, we will see how to build automated tests for your services as integrated into ABP Framework and other infrastructure components. We will start by understanding ABP integration, how the database is used in integration tests, and how to create initial test data. Then, we will write example tests for repositories, domain, and application services. Let's start with ABP integration.

Understanding ABP integration

ABP provides the `Volo.Abp.TestBase` NuGet package, which includes the `AbpIntegratedTest<TStartupModule>` base class for our integration tests. We can inherit from that class to write tests as completely integrated to ABP Framework. The following example shows the main parts of such a test class:

```
public class SampleTestClass
    : AbpIntegratedTest<MyTestModule>
{
    private IMyService _myService;

    public SampleTestClass()
    {
        _myService = GetRequiredService<IMyService>();
```

```
    }

    [Fact]
    public async Task TestMethod()
    {
        await _myService.DoItAsync();
    }
}
```

In this example, I've inherited from the `AbpIntegratedTest<MyTestModule>` class, where `MyTestModule` is my startup module class. `MyTestModule` should depend on `AbpTestBaseModule`, as shown in the following example:

```
[DependsOn(typeof(AbpTestBaseModule))]
public class MyTestModule : AbpModule
{

}
```

In the constructor of `SampleTestClass`, I've resolved an example service using the `GetRequiredService` method and assigned it to a class field. We could resolve a service from the DI system since all the infrastructure is available, just like at runtime. I don't need to care about the dependencies of the service. Finally, I called a method of the example service in my test method.

While writing integration tests is that simple, test projects in the startup template have a little more. Check the `EventHubTestBaseModule` class (in the `EventHub.TestBase` project of the EventHub solution). You will see that it is disabling background jobs and authorization, seeding some test data, and doing other configurations.

We've learned the basics of integrating with ABP in our test classes. In the next section, you will learn how to deal with the database in tests.

Mocking the database

The database is one of the most fundamental aspects when you are building integration tests. Assume that you are using SQL Server in your solution. Using a real SQL Server database has some fundamental problems; your tests affect each other since they will work on the same database. A test's change in the database may break subsequent tests. You may not run tests in parallel. Test execution speed will be slow since your application will communicate to SQL Server as an external process. No need to mention that SQL Server should be installed and available in your test environment.

EF Core provides an in-memory database option, but it is very limited. For example, it has no transaction support and cannot execute SQL commands. So, I don't suggest using it at all.

The ABP startup template has been configured to use SQLite in-memory database for EF Core (it also uses an in-memory database for MongoDB using the `Mongo2Go` library). SQLite is a real relational database management system and will be sufficient for most of the applications.

Check the `EventHubEntityFrameworkCoreTestModule` class (in the `EventHub.EntityFrameworkCore.Tests` project of the EventHub solution) to see the SQLite setup. It creates a separate in-memory SQLite database for each test case, creates tables inside the database, and seeds the test data. In this way, every test method starts with the same initial state and doesn't affect other tests. We will see seeding the test data in the next section.

Seeding the test data

Writing tests against an empty database is not so practical. Assume that you want to query events in the database or want to test whether the event registration code works. You first need to insert some entities into the database. It can be tedious to prepare the database for each test. Instead, we can create some initial entities in the database that are available for each test.

The ABP startup solution template uses ABP's data seeding system to fill some initial data into the database. See the `EventHubTestDataSeedContributor` class (in the `EventHub.TestBase` project of the EventHub solution). It creates some users, organizations, and events in the database, so we can directly write tests assuming that the initial data exists.

We've talked about ABP's integration test infrastructure, mocking and seeding the database. Now, we can write some integration tests, starting from the repositories.

Testing repositories

Let's see the `EventRegistrationRepository_Tests` class (in the `EventHub.Domain.Tests` project of the EventHub solution) as an example:

```
public class EventRegistrationRepository_Tests
    : EventHubDomainTestBase
{
    private readonly IEventRegistrationRepository
```

```
            _repository;
    private readonly EventHubTestData _testData;


    public EventRegistrationRepository_Tests()
    {
        _repository = GetRequiredService<
                            IEventRegistrationRepository>();
        _testData = GetRequiredService<EventHubTestData>();
    }
    // TODO: Test methods come here...
}
```

This class inherits the EventHubDomainTestBase class, which is indirectly inherited from the AbpIntegratedTest<T> class we've explored in the *Understanding the ABP integration* section. So, in the constructor, we could resolve the IEventRegistrationRepository and EventHubTestData services from the DI system. You can investigate the EventHubTestData class yourself (in the EventHub.TestBase project of the EventHub solution). It basically stores the Id values of the entities that are initially seeded into the database to reach them in the tests.

Let's see the first test method of the EventRegistrationRepository_Tests class. Here it is:

```
[Fact]
public async Task
    Exists_Should_Return_False_If_Not_Registered()
{
    var exists = await _repository.ExistsAsync(
        _testData.AbpMicroservicesFutureEventId,
        _testData.UserJohnId);
    exists.ShouldBeFalse();
}
```

This test simply executes the ExistsAsync method and checks the result to be false. It should return false because we know the user John has not registered to the given event. We know that because we've written the initial data in the database (see the *Seeding the test data* section). Let's write another test, as follows:

```
[Fact]
public async Task Exists_Should_Return_True_If_Registered()
```

```
{
    await _repository.InsertAsync(
        new EventRegistration(
            Guid.NewGuid(),
            _testData.AbpMicroservicesFutureEventId,
            _testData.UserJohnId));

    var exists = await _repository.ExistsAsync(
        _testData.AbpMicroservicesFutureEventId,
        _testData.UserJohnId);
    exists.ShouldBeTrue();
}
```

This time, we are creating the registration record in the database, so we expect the same `ExistsAsync` call to return `true`. In this way, we can prepare the database for a particular test to get the expected result.

ABP's repositories provide the `GetQueryableAsync` method, so we can directly use **Language-Integrated Query** (**LINQ**) on the database. See the following example test method (this test is not included in EventHub but is provided here to be an example of using `queryable` in tests):

```
[Fact]
public async Task Test_Querying()
{
    var queryable = await _repository.GetQueryableAsync();
    var exists = await queryable.Where(
        x => x.EventId ==
            _testData.AbpMicroservicesFutureEventId &&
            x.UserId == _testData.UserJohnId
        ).FirstOrDefaultAsync();
    exists.ShouldBeNull();
}
```

This method queries the same registration using the `Where` and `FirstOrDefaultAsync` LINQ extension methods. If you try to run this test, you will see it throws an exception (of type `ObjectDisposedException`) because the `GetQueryableAsync` method requires an active **unit of work (UoW)** (see the *Understanding the UoW system* section of *Chapter 6, Working with the Data Access Infrastructure,* to remember ABP's UoW system). The base test class provides the `WithUnitOfWorkAsync` method to execute code in a UoW, so we can fix the test code as shown in the following code block:

```
[Fact]
public async Task Test_Querying_With_Uow()
{
    await WithUnitOfWorkAsync(async () =>
    {
        var queryable =
            await _repository.GetQueryableAsync();
        var exists = await queryable.Where(
            x => x.EventId ==
                _testData.AbpMicroservicesFutureEventId &&
                x.UserId == _testData.UserJohnId
        ).FirstOrDefaultAsync();
        exists.ShouldBeNull();
    });
}
```

You can see the source code of the `WithUnitOfWorkAsync` method. It just uses `IUnitOfWorkManager` to create a UoW scope.

We've created some test methods for repositories. You can test any service (that was registered to the DI system) in the same way. I will show some example tests for domain and application services in the next two sections.

Testing domain services

Testing domain services is similar to testing repositories since you should also care about the UoW for domain services. The following code block shows an example test case from the EventManager_Tests class (in the EventHub.Domain.Tests project of the EventHub solution):

```
[Fact]
public async Task Should_Update_The_Event_Capacity()
{
    const int newCapacity = 42;
    await WithUnitOfWorkAsync(async () =>
    {
        var @event = await _eventRepository.GetAsync(
            _testData.AbpMicroservicesFutureEventId);
        await _eventManager.SetCapacityAsync(
            @event,
            newCapacity
        );
    });

    var @event = await _eventRepository.GetAsync(
        _testData.AbpMicroservicesFutureEventId);
    @event.Capacity.ShouldBe(newCapacity);
}
```

The test's purpose is to increase the capacity of an event using the EventManager domain service and see whether it works. It uses the WithUnitOfWorkAsync method to call the SetCapacityAsync method because the SetCapacityAsync method internally executes the CountAsync LINQ extension method and requires an active UoW. If you don't want to check the domain service's internals in every case, I suggest always starting a UoW while using domain services or repositories in your tests. After the UoW, I've re-queried the same event from the database to check whether the capacity has been updated.

You can explore the EventManager_Tests class and other test classes inside the EventHub.Domain.Tests project for all details and more complex test cases. In the next section, I will show testing application services.

Testing application services

In this section, we will examine one more test, written for the
`EventRegistrationAppService` class (defined in the `EventHub.Application`
project of the EventHub solution). `EventRegistrationAppService_Tests`
(defined in the `EventHub.Application.Tests` project of the EventHub solution)
is the test class that contains tests for that application service. You can explore the class
inside the solution. Here, I will show it partially to explain how it works.

Let's start with the test method for registering an event by the current user. You can see
this here:

```
[Fact]
public async Task Should_Register_To_An_Event()
{
    Login(_testData.UserAdminId);

    await _eventRegistrationAppService.RegisterAsync(
        _testData.AbpMicroservicesFutureEventId
    );

    var registration = await GetRegistrationOrNull(
        _testData.AbpMicroservicesFutureEventId,
        _currentUser.GetId()
    );
    registration.ShouldNotBeNull();
}
```

The first line sets the current user to the admin user, which is needed because the
`EventRegistrationAppService.RegisterAsync` method works for the current
user. Let's see how the `Login` method was implemented, as follows:

```
private void Login(Guid userId)
{
    _currentUser.Id.Returns(userId);
    _currentUser.IsAuthenticated.Returns(true);
}
```

It configures the _currentUser object to return the given userId value when we use its Id property. As you may guess, _currentUser is a mock (fake) object of type ICurrentUser. The mock object is configured in the AfterAddApplication method, as illustrated in the following code snippet:

```
protected override void AfterAddApplication(
    IServiceCollection services)
{
    _currentUser = Substitute.For<ICurrentUser>();
    services.AddSingleton(_currentUser);
}
```

This method overrides the AfterAddApplication method of the AbpIntegratedTest<T> base class. We can override this method to make a last touch to the DI configuration before the initialization phase completes. Here, I've created a mock object using the NSubstitute library and added the object as a singleton service (remember that the last registered class/object is used for a service). In this way, I can change its value, and all the services using ICurrentUser are affected.

After setting the current user, the test method calls the EventRegistrationAppService.RegisterAsync method as you normally do. Finally, I checked the database to see whether a registration record was saved. The GetRegistrationOrNull method's implementation is shown in the following code block:

```
private async Task<EventRegistration>
    GetRegistrationOrNull(Guid eventId, Guid userId)
{
    return await WithUnitOfWorkAsync(async () =>
    {
        return await _eventRegistrationRepository
            .FirstOrDefaultAsync(
                x => x.EventId == eventId && x.UserId ==
                    userId
            );
    });
}
```

I've used `WithUnitOfWorkAsync` here again because the `FirstOrDefaultAsync` method requires an active UoW.

As we've seen in the examples, writing integration tests is easy and mostly straightforward with ABP Framework. We rarely need to mock services and deal with the dependencies of the service we are targeting for the test.

Integration tests run slower than unit tests, but they allow you to test the integration between components and additionally test database queries in a way you can't with unit tests. I suggest going balanced and pragmatic—build both unit and integration tests for your solutions.

Summary

Preparing tests is an essential practice for building any kind of software solution. As we've seen in this chapter, ABP provides the fundamental infrastructure to help you write tests for your applications.

We have explored unit and integration testing with ABP Framework with examples. I've selected examples from the EventHub solution. That solution also contains more complex tests, and I suggest you explore them.

By now, you should be writing automated tests to cover your server-side code. You've seen how the ABP startup solution is structured and how the database was mocked. You've learned how to deal with exceptions, UoWs, data seeding, object mocking, and other common test patterns.

This was the last chapter of the book. If you've read so far and followed the examples, you've learned the fundamentals, features, and best practices of using ABP Framework. You are more than ready to build your ABP based solutions to realize your software ideas.

You can refer to this book on your development journey, and also check ABP Framework's documentation on `https://docs.abp.io` whenever you need more details and up-to-date information.

Lastly, feel free to create issues on ABP Framework's GitHub repository if you have any problem: `https://github.com/abpframework/abp`. I will continue to be one of the active contributors of this great project and try to answer your questions.

I am Halil İbrahim Kalkan, author of Mastering ABP Framework. I really hope you enjoyed reading this book and found it useful for increasing your productivity and efficiency in ABP Framework.

It would really help me (and other potential readers!) if you could leave a review on Amazon sharing your thoughts on Mastering ABP Framework here.

Go to the link below or scan the QR code to leave your review:

`https://packt.link/r/1801079242`

Your review will help me to understand what's worked well in this book, and what could be improved upon for future editions, so it really is appreciated.

Best Wishes,

Halil İbrahim Kalkan

Index

B

`Packt.com`

Subscribe to our online digital library for full access to over 7,000 books and videos, as well as industry leading tools to help you plan your personal development and advance your career. For more information, please visit our website.

Why subscribe?

- Spend less time learning and more time coding with practical eBooks and Videos from over 4,000 industry professionals

- Improve your learning with Skill Plans built especially for you

- Get a free eBook or video every month

- Fully searchable for easy access to vital information

- Copy and paste, print, and bookmark content

Did you know that Packt offers eBook versions of every book published, with PDF and ePub files available? You can upgrade to the eBook version at `packt.com` and as a print book customer, you are entitled to a discount on the eBook copy. Get in touch with us at `customercare@packtpub.com` for more details.

At `www.packt.com`, you can also read a collection of free technical articles, sign up for a range of free newsletters, and receive exclusive discounts and offers on Packt books and eBooks.

Other Books You May Enjoy

If you enjoyed this book, you may be interested in these other books by Packt:

Web Development with Blazor

Jimmy Engström

ISBN: 978-1-80020-872-8

Understand the different technologies that can be used with Blazor, such as Blazor Server and Blazor WebAssembly.

- Find out how to build simple and advanced Blazor components.
- Explore the differences between Blazor Server and Blazor WebAssembly projects
- Discover how Entity Framework works and build a simple API.
- Get up to speed with components and find out how to create basic and advanced components.

Customizing ASP.NET Core 6.0

Jürgen Gutsch

ISBN: 978-1-80323-360-4

- Explore various application configurations and providers in ASP.NET Core 6.
- Enable and work with caches to improve the performance of your application.
- Understand dependency injection in .NET and learn how to add third-party DI containers.
- Discover the concept of middleware and write your middleware for ASP.NET Core apps.
- Create various API output formats in your API-driven projects.

Packt is searching for authors like you

If you're interested in becoming an author for Packt, please visit `authors.packtpub.com` and apply today. We have worked with thousands of developers and tech professionals, just like you, to help them share their insight with the global tech community. You can make a general application, apply for a specific hot topic that we are recruiting an author for, or submit your own idea.